AMS SHORT COURSE LECTURE NOTES
Introductory Survey Lectures
A Subseries of Proceedings of Symposia in Applied Mathematics

Volume 21 MATHEMATICAL ASPECTS OF PRODUCTION AND DISTRIBUTION OF ENERGY
Edited by P. D. Lax (San Antonio, Texas, January 1976)

Volume 22 NUMERICAL ANALYSIS
Edited by G. H. Golub and J. Oliger (Atlanta, Georgia, January 1978)

Volume 23 MODERN STATISTICS: METHODS AND APPLICATIONS
Edited by R. V. Hogg (San Antonio, Texas, January 1980)

Volume 24 GAME THEORY AND ITS APPLICATIONS
Edited by W. F. Lucas (Biloxi, Mississippi, January 1979)

Volume 25 OPERATIONS RESEARCH: MATHEMATICS AND MODELS
Edited by S. I. Gass (Duluth, Minnesota, August 1979)

Volume 26 THE MATHEMATICS OF NETWORKS
Edited by S. A. Burr (Pittsburgh, Pennsylvania, August 1981)

Volume 27 COMPUTED TOMOGRAPHY
Edited by L. A. Shepp (Cincinnati, Ohio, January 1982)

Volume 28 STATISTICAL DATA ANALYSIS
Edited by R. Gnanadesikan (Toronto, Ontario, August 1982)

Volume 29 APPLIED CRYPTOLOGY, CRYPTOGRAPHIC PROTOCOLS, AND COMPUTER SECURITY MODELS
By R. A. DeMillo, G. I. Davida, D. P. Dobkin, M. A. Harrison, and R. J. Lipton (San Francisco, California, January 1981)

Volume 30 POPULATION BIOLOGY
Edited by Simon A. Levin (Albany, New York, August 1983)

Volume 31 COMPUTER COMMUNICATIONS
Edited by B. Gopinath (Denver, Colorado, January 1983)

Volume 32 ENVIRONMENTAL AND NATURAL RESOURCE MATHEMATICS
Edited by R. W. McKelvey (Eugene, Oregon, August 1984)

Volume 33 FAIR ALLOCATION
Edited by H. Peyton Young (Anaheim, California, January 1985)

AMS SHORT COURSE LECTURE NOTES

Introductory Survey Lectures

published as a subseries of

Proceedings of Symposia in Applied Mathematics

PROCEEDINGS OF SYMPOSIA
IN APPLIED MATHEMATICS
Volume 34

Mathematics of Information Processing

Michael Anshel and
William Gewirtz, Editors

American Mathematical Society
Providence, Rhode Island

LECTURE NOTES PREPARED FOR THE
AMERICAN MATHEMATICAL SOCIETY SHORT COURSE

MATHEMATICS OF INFORMATION PROCESSING

HELD IN LOUISVILLE, KENTUCKY
JANUARY 23–24, 1984

The AMS Short Course Series is sponsored by the Society's Committee on Employment and Education Policy (CEEP). The series is under the direction of the Short Course Advisory Subcommittee of CEEP.

Library of Congress Cataloging in Publication Data
Main entry under title:
Mathematics of information processing.
(Proceedings of symposia in applied mathematics, ISSN 0160-7634; v. 34)
Bibliography: p.
Contents: Diameters of communication networks / F. R. K. Chung–The theory of data dependencies / Ronald Fagin and Moshe Y. Vardi–Transaction management / Hector Garcia-Molina–[etc.] 1. Electronic data processing–Mathematics–Congresses. I. Anshel, Michael, 1941– . II. Gewirtz, William, 1948– . III. Series.
QA76.9.M35M39 1986 005.7 85-26693
ISBN 0-8218-0086-8 (alk. paper)

1980 *Mathematics Subject Classification* (1985 *Revision*). Primary 68-06.

Printed in the United States of America.
This volume was printed directly from copy prepared by the authors.
The paper used in this book is acid-free and falls within the guidelines
established to ensure permanence and durability.

CONTENTS

CONTRIBUTORS

F. R. CHUNG, Bell Laboratories, Murray Hill, New Jersey.

RONALD FAGIN, IBM Research Laboratory, San Jose, California.

HECTOR GARCIA-MOLINA, Department of Electrical Engineering and Computer Science, Princeton University, Princeton, New Jersey.

BARRY E. JACOBS, Department of Computer Science, University of Maryland, College Park, Maryland and National Aeronautics and Space Administration, Goddard Space Flight Center, Greenbelt, Maryland.

VICTOR S. MILLER, IBM Watson Research Center, Yorktown Heights, New York.

AUGUSTIN A. TUZHILIN, Department of Computer Science, College of Staten Island, Staten Island, New York.

MOSHE E. VARDI, IBM Research Laboratory, San Jose, California.

PREFACE

This volume contains the lecture notes prepared by six speakers for the American Mathematical Society Short Course on the Mathematics of Information Processing given in Louisville, Kentucky, January 23-24, 1984. The Short Course Advisory Subcommittee of the AMS approved this concept and recommended publication of these lecture notes.

The Mathematics of Information Processing is not a single topic but rather a collection of methodologies whose end-goal is the creation of automated information systems.

The viewpoint represented here is largely that of American researchers with heavy emphasis on the mathematical problems of database systems and communication networks. This reflects the rapid introduction of the products of information technology in the workplace and in the home. By way of contrast a systems-theoretic approach developed in the Soviet Union is included which also provides a self-contained background should the reader care to probe more deeply into the subject.

The idea for the Short Course arose while one of us (Anshel) was a NASA-ASEE Summer Faculty Fellow at Goddard Space Flight Center, NASA. There we learned of a proposed futuristic information system, IESIS (Intelligent Earth Sensing Information System). The real-world implementation of the Short Course rested on the shoulders of the practitioner among us (Gewirtz) who also chaired a lively panel discussion with a surprise guest (Bob Targan).

We would like to thank the speakers for their efforts in the preparation of these lectures and their patience concerning the idiosyncracies of the co-director.

Finally we also wish to thank Stefan Burr who acted as wise counsel and enthusiastic supporter.

Michael Anshel
City College of the
City University of New York

William Gewirtz
ATT Communications
Basking Ridge, New Jersey

Proceedings of Symposia in Applied Mathematics
Volume 34, 1986

Diameters Of Communication Networks

F. R. K. CHUNG*

ABSTRACT. When graphs are used to model the linkage structure of communication networks, the diameter of the graph corresponds to the maximum number of links over which a message between two nodes must travel. In cases where the number of links in a path is roughly proportional to the time delay or signal degradation encountered by messages sent along the path, the diameter is then involved in the complexity analysis for the performance of the networks. A variety of interrelated diameter problems will be discussed here, including: determining extremal graphs of bounded degrees and small diameters, finding orientations for undirected or mixed graphs to minimize diameters; investigating diameter bounds for networks with possible node and link failures, and algorithms aspects for determining the diameters of graphs.

1. Introduction. Modern communication networks typically are highly complex structures formed by various interconnected components. Many principal characteristics of communication networks often result from the topology of the underlying connection patterns of the network. Graph theory can then be used to study the linkage structure of the network and to model problems arising in the optimization and analysis of the networks, (see [11]).

A graph G consists of a finite set $V(G)$ of vertices (or nodes) together with a prescribed set $E(G)$ of unordered pairs of vertices of $V(G)$. The vertices represent objects in a network and the pairs, called edges (or links) represent the interconnections between objects. We note that the exact geometric positions of the vertices or the lengths of the edges are not important unless specified.

---

* 1980 Mathematics Subject Classification 05, 94C15.
Bell Communications Research, Morristown, New Jersey 07960.

0160-7634/86 $1.00 + $.25 per page

For two vertices u and v in a graph G, a path P of length t from u to v is a sequence of distinct vertices $u = a_0, a_1, \ldots, a_t = v$, together with edges $\{a_i, a_i+1\}, i = 0, \ldots, t-1$, *in* G. A graph is said to be *connected* if every pair of vertices are joined by a path. In a connected graph G, the distance $d_G(u,v)$ between two vertices u and v is the length of a shortest path joining u and v in G. The diameter $D(G)$ *of* G is the maximum value of $d_G(u,v)$, taken over all pairs of vertices $u, v \epsilon V(G)$ (see Fig. 1).

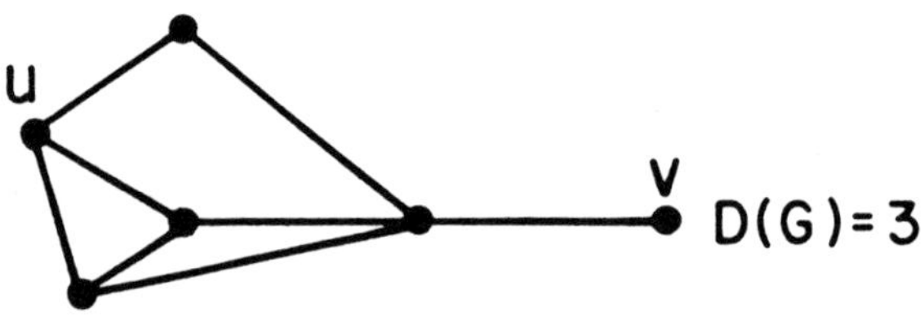

FIGURE 1

In the graph model for communication networks, the diameter of the graph corresponds to the maximum number of edges over which a message between two nodes must travel. In cases where the number of edges in a path is roughly proportional to the time delay or signal degradation encountered by messages sent along the path, the diameter is directly involved in the analysis and the optimization of the networks. In particular, diameter-related problems often arise in connection with analyzing the computational complexity of routing, distributing and scheduling algorithms.

Before we proceed to several interrelated diameter problems, we will first introduce some definitions.*

For a given graph G, a subgraph G' *of* G is given by taking $V(G') \subseteq V(G)$ and $E(G') \subseteq E(G)$. A maximum connected subgraph of G is called a *connected component of* G. A *bridge* of a connected graph G is an edge whose removal disconnects G. If G has no bridge, it is called bridgeless.

In a graph G, a path is formed by a sequence of distinct vertices $a_0, a_1, \ldots, a_t$ together with edges $\{a_i, a_i+1\}$, $i = 0,\ldots,t-1$. A cycle is formed by a path $a_0, \ldots, a_t$ together with the edge $\{a_t, a_0\}$. An acyclic

* For undefined graph-theoretic terminology, the reader is referred to [2, 15, 33].

connected graph is called a *tree*. For a vertex v in G the degree of v, denoted by $deg_G(v)$, is the number of edges $\{u,v\}$ in G which contain v. These u's are called the neighbors of v. We note that the maximum degree of a graph is also a useful parameter in network optimization since practical networks commonly satisfy some degree constraints. We will discuss extremal graphs with small diameter and bounded degrees in Sections II and III.

In cases that some links only allow one-way traffic in communication networks we will then consider directed graphs or mixed graphs. A directed graph G is formed by a vertex set together with a prescribed set of ordered pairs of vertices. The edge set of a mixed graph contains ordered and unordered pairs of vertices. Note that both undirected and directed graphs are special cases of mixed graphs. We shall denote an edge with end vertices u and v,, by $[u,v]$ if it is either undirected or directed from u to v. Similarly we define paths and connectivity in mixed graphs. A path P in G from u to v is a sequence of distinct vertices $u = a_0, a_1, \ldots, a_t = v$ so that $[a_i, a_{i+1}], i = 0,...,t-1$, are edges in G, and P is called a walk if a_i's are not necessarily distinct. G is strongly connected if every pair of vertices is joined by a path. An orientation of a mixed graph G is any directed graph obtained by directing every undirected edge in G. G is orientable if there is an orientation of G which is strongly connected. In Section IV we will discuss the problem of choosing orientations of a mixed graph subject to minimizing the diameter in the resulting directed graphs.

In Section V we will consider diameter problems associated with networks having possible link or node failures. We will study extremal problems for fault-tolerant graphs with small diameters and investigate the diameter bounds in general graphs while a small number of vertices or edges are deleted. In Section VI we will discuss fast algorithms for finding the diameter of a graph. Many unresolved problems will be mentioned.

2. Large graphs with bounded degree and diameter.

The following problem has been studied by many researchers in the past (see [3, 4, 5, 6]):

How many vertices can a graph have which has diameter D and degree at most k?

This problem can be viewed as a network optimization problem of connecting as many processors as possible while each processor has a bounded number of ports (degree constraint) and the delay in data transmission is small (diameter constraint). The maximum number $n(k,D)$ of vertices in a graph with diameter D and maximum degree k is bounded above by

$$n(k,D) \leq 1+k+ \cdots +k(k-1)^{D-1} = n_0(k,D)$$

since there are at most $k(k-1)^{i-1}$ vertices at a distance $i \geq 1$ from a vertex. The upper bound $n_0(k,D)$, called Moore bound, is provably unreachable [34] for almost all nontrivial values of k and D. The only graphs, called Moore graphs, which achieve the Moore bounds must be one of the following [34]:

(i) $D = 1$, $(k+1)$-cliques

(ii) $k = 2$, $(2D+1)$-cycles

(iii) $D = 2$ and $k = 3$, the Petersen graph

(iv) $D = 2$ and $k = 7$, the Hoffman-Singleton graph

(v) (possibly) $D = 2$ and $k = 57$.

Problem 1 [34]: Is there a Moore graph of diameter 2 and degree 57?
Best known lower bound for $n(k,D), k,D \leq 10$ can be found in Table 1 (see [7]).

D / k	2	3	4	5	6	7	8	9	10
3	10	20	38	70	128	180	286	462	708
4	15	40	95	364	731	856	1872	3708	7100
5	24	66	174	532	2734	2988	7600	11340	30240
6	32	105	320	820	7817	12285	21840	65520	185640
7	50	122	480	1550	10546	35154	78120	281232	1218672
8	57	200	807	2652	39223	66402	175770	749952	3656016
9	74	585	1248	5150	74906	215688	588240	2941200	15686400
10	91	650	1755	8400	132869	486837	1348164	7489800	47059200
11	94	715	3200	12285	156864	863590	2790060	16852050	120969030
12	133	780	4680	18860	354323	1527890	4782960	36270780	326835600
13	136	845	6560	33345	531440	2657340	9920736	72541560	581071680
14	183	910	8200	42705	804481	4783212	18601380	145090764	1556138804
15	186	1215	11712	55062	1417248	6837978	28960848	282740976	2355482304
16	197	1600	14640	132496	1771560	11664786	54301590	481474098	5743901520

TABLE 1

For the upper bound the only result beyond $n(k,D) < n_0(k,D)$ (except for (i)-(v)) was obtained by P. Erdős, S. Fajtlowicz and A. J. Hoffman [27] who proved that

$$n(2k,2) \leq n_0(2k,2) - 2 \text{ for } k > 1.$$

Problem 2 [12, 26]: Is it true that for every integer c there exist k and D such that

$$n(k,D) \leq n_0(k,D) - c \ ?$$

There are two different approaches for establishing the lower bounds for $n(k,D)$: explicitly construct such a good graph or prove by probabilistic methods the existence of a good graph. For practical concern the first approach is much more desirable although the second often gives better bounds (so far).

Many of the explicit constructions are extensions or modifications of the de Bruijn graphs. Here we describe briefly the structure of de Bruijn graphs.

For given integers r and s and $r \geq 3$, the de Bruijn graph $B(r,s)$ has s^r vertices represented by r-tuples $(a_1, a_2, \ldots, a_r)$, where $a_i \epsilon \{1,\ldots,s\}$ and $(a_1, a_2, \ldots, a_r)$ is adjacent to $(a_2, \ldots, a_r, b)$ and $(b, a_1, \ldots, a_{r-1})$ for any $b \epsilon \{1,\ldots,s\}$. It is easily seen that $B(r,s)$ has diameter r and degree $2s$. This gives

$$\mu_D = \limsup_{k\to\infty} \frac{n(k,D)}{n_0(k,D)} \geq 2^{-D}.$$

For some small fixed values of D, the ratio of $n(k,D)$ and $n_0(k,D)$ can be arbitrarily close to one for sufficiently large k due to the explicit construction of large classes of graphs using combinational structures such as generalized n-gons and product constructions (see [2,5]), which we will describe.

A generalized polygon can be viewed as a bipartite graph G with vertex set $P \cup L$ having the property that for any two vertices x,y of distance $d(x,y) <$ diameter (G), there is a unique path of length $d(x,y)$ joining x and y. Elements in P and L will be called points and lines, respectively, and we say $x \epsilon P$ belongs to (or is incident with) $l \epsilon L$ if and only if (x,l) is an edge in G.

If the degree of each vertex of G is at least 3 and the diameter is n, G is called a thick generalized n-gon. It turns out that any two vertices in P have the same degree and any two vertices in L have the same degree. We say the generalized n-gon has parameter (s,t) if every line contains $s+1$ points and every point is contained in $t+1$ lines. A theorem of Feit and Higman [29] states that thick generalized n-gons only exist for $n = 2,3,4,6$ and 8.

For example, the graphs of generalized 2-gons are precisely the complete bipartite graphs. The thick generalized 3-gons are the nondegenerate projective planes with q^2+q+1 points and lines where q is a prime power (see [36]).

Generalized 4-, 6-, and 8-gons are known as generalized quadrangles, hexagons, and octagons, respectively. There are several types of generalized quadrangles, namely, the classical polar spaces Sp (4,q), $0^{-}(6,q)$ and $U(5,q^2)$ where q is a prime power. These constructions are rather complicated (the reader is referred to [36,48,49]). There are two known types of generalized hexagons $G_2(q)$ and ${}^3D_4(q^3)$ [see 36,48] and one generalized octagon ${}^2F_4(q)$ which requires q to be an odd power of 2 (see [36,52]).

Now for a bipartite graph G with vertex set $P \cup L$, the self product G^2 has vertex set $P \times L = \{(p,l): p \epsilon P, l \epsilon L\}$ while (p,ℓ) is adjacent to (p',ℓ') if p is contained in ℓ' and p' is contained in ℓ. It is not difficult to check that G^2 has diameter one less than the diameter of G and G is regular of degree k_2k_2 if vertices in P have degree k_1 and vertices in L have degree k_2 in G (see [5]).

Now suppose we consider the generalized octagon ${}^2F_4(q)$ which has diameter 8 with parameter (q,q^2) and contains $(1+o(1))q^{10}$ points and $(1+o(1))q^{11}$ lines. The self product of ${}^2F_4(q)$ has then $(1+o(1))q^{21}$ vertices with diameter 7 and maximum degree $(1+o(1))q^3$. As an immediate consequence, we have

$$\mu_7 = \lim_{k\to\infty} sup \frac{n(k,7)}{n_0(k,7)} = 1.$$

Although these methods were known in the proofs of $\mu_3 = \mu_5 = 1$. Bert Wells first proved this fact about μ_7. Also, he pointed out that by using the following inequality derived by Delorme in [24]

$$\mu_{D+1} \geq 2\mu_D D^D (D+1)^{-(D+1)},$$

one gets $\mu_8 \geq 2{\cdot}7^7 8^{-8}, \mu_9 \geq 4{\cdot}7^7 9^{-9}$ and $\mu_{10} \geq 8{\cdot}7^7 10^{-10}$.

Here are best known lower bounds for μ_D, for $D \leq 10$,

D	1	2	3	4	5	6	7	8	9	10
μ_D	1	1	1	$3^3\, 2^{-7}$	1	$2{\cdot}5^5\, 6^{-6}$	1	$2{\cdot}7^7\, 8^{-8}$	$4{\cdot}7^7\, 9^{-9}$	$5{\cdot}2^{-10}$

Using probabilistic approaches B. Bollobas and W. F. de la Vega [14] obtained a much stronger lower bound: $n(k,D) \geq \dfrac{c}{k^3 D \log (k-1)} n_0(k,D)$ for some constant c.

The directed analogue of this problem turns out to be much more tractable. We can define the maximum number $\bar{n}(k,D)$ of vertices in a directed graph of outdegree $\leq k$ and diameter D. Then $\bar{n}(k,D)$ must satisfy

$$\bar{n}(k,D) < 1+k+k^2+...+k^D = \bar{n}_0(k,D).$$

In 1974, Plesnik and Znam [43] showed that the directed Moore bound $\bar{n}_0(k,D)$ is not achievable for $k,D > 1$, which was rediscovered later in [16]. On the other hand, Kautz in 1968 (see [16]) gave a construction of a directed graph on $k^D + k^{D-1}$ vertices with outdegree k and diameter D, which again was rediscovered in [44].

Problem 3 [16]: Determine the exact values for $\bar{n}(k,D)$.

3. Sparse graphs with bounded degree and diameter.

P. Erdös and A. Rényi [25] asked the following question in 1962:
Suppose there are n cities such that the airport of each city can handle at most k flights. It is desirable to schedule the flights in such a way that from each city it is possible to fly to another city with at most t stops along the way. What is the minimum number of flights which must be set up to satisfy the stated requirements? In other words, what is the least number $e(n,k,D)$ of edges in a graph with n vertices having degree at most k and diameter at most D where D = t+1?

The above simple-looking problem turns out to be unexpectly difficult. Relatively few exact values for $e(n,k,D)$ are known so far except for some of the cases with $D \leq 3$. Some partial results and estimates were obtained in the past [12, 25, 28] and numerous questions still remain unresolved.

In Fig. 2 we illustrate the extremal graph G with n vertices having diameter 2, degree $\leq k$,$(2n-2)/3 \leq k \leq n-5$, and $e(n,k,2) = 2n-4$ edges (see [12,25]).

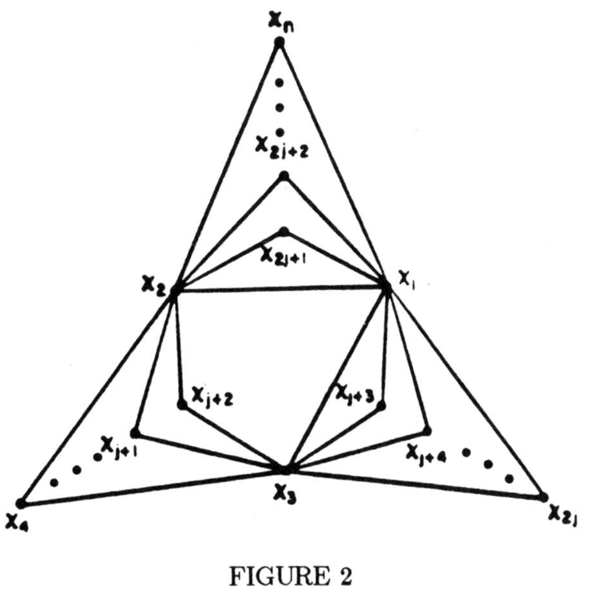

FIGURE 2

Known exact values for $e(n,k,d)$ are (see [25, 26], also see [12]):

$$e(n,\ k,\ 2) = \begin{cases} n-1 & \text{if } k = n-1 \\ n+k-2 & \text{if } k = n-2 \text{ or } n-3 \\ 2n-5 & \text{if } k = n-4 \\ 2n-4 & \text{if } (2n-2)/3 \le k \le n-5 \\ 3n-k-6 & \text{if } \frac{3n-3}{5} \le k < \frac{2n-2}{3} \\ 5n-4k-10 & \text{if } \frac{5n-3}{9} \le k \le \frac{3n-3}{5} \\ 4n-2k-11 & \text{if } \frac{n+1}{2} \le k < \frac{5n-3}{9} \end{cases}$$

$e(n,k,3)=n+\binom{s}{2}-1$ if $[n/(s+1)]+s-1 \le k < [n/s]+s-2$ and $1 \le s \le [(n/2)^{1/3}]$.

J. Pach and L. Suranyi [40] proved that $e(n,cn,2)/n$ tends to a limit $g(c)$ for n large and any fixed c between 0 and 1. The function $g(c)$ is a piecewise linear function except at a sequence of "turning" points. The values of $g(c)$ can be determined using linear programming for any fixed c.

For $D \ge 3$, a lower bound for $e(n,k,D)$ is obtained in [26]:

$$e(n,k,D) \ge \frac{n^2}{k^{D-1}}(1-(n/k^D)^{1/3}).$$

It is also proved in [22] that

$$e(n+p,k,D+2) \le e(n,k,D)+p$$

for $0 \le p \le kn-2e(n,k,D)$.

Problem 4: What is the least number $\bar{e}(n,k,D)$ of edges in a directed graph with n vertices having diameter D and degree (the sum of indegree and outdegree) $\le k$?

Clearly, $\bar{e}(n,k,D) \le 2e(n,[k/2],D)$.

G. Katona and E. Szemeredi [36] proved that any directed graph with n vertices, which does not contain any cycle of length 2 and has diameter 2, must have at least $n \log_2 n$ edges and that is the best possible. We can ask analogous questions for directed graphs with degree constraints and other constraints as well, for example, having no small cycles.

4. Orientations of mixed graphs with small diameters.

In 1939 H. Robbins [45] asked the following question: "When is it possible to find an assignment of one-way directions for all the streets in a town while preserving the property that it is possible to reach any point in town from any other point?".

He solved [45] this problem by proving an (undirected) graph is orientable if and only if it is connected and has no bridge. F. Boesch and R. Tindell [8] considered the more general case for a town in which some, but not all, of the streets are already one-way streets. They proved [8] that a mixed graph is orientable if and only if it is strongly connected and has no bridge.

J. A. Bondy and U. S. R. Murty raised the problem of determining how much the diameter can increase in the process of orienting edges while preserving strong connectivity. V. Chvátal and C. Thomassen [22] subsequently proved that every bridgeless (undirected) graph of diameter D admits an orientation of diameter $2D^2+2D$. On the other hand, they show there is a bridgeless graph G of diameter D with the property that any orientation of G has diameter at least $D^2/4+D$ (see Fig. 3 for the case of $D=4$).

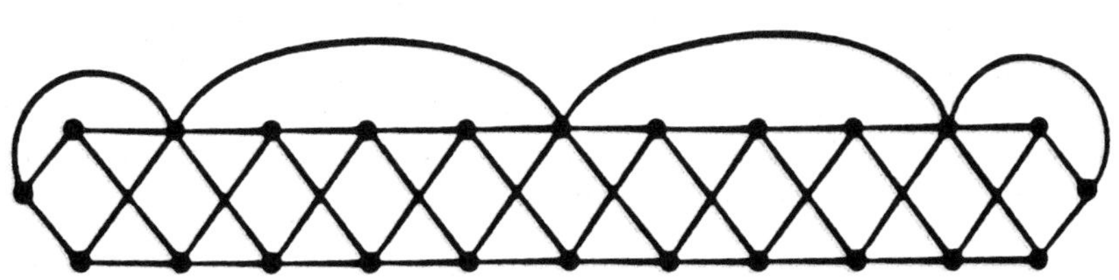

Figure 3

Let $f(D)$ denote the least number such that any bridgeless graph of diameter D admits an orientation of diameter $\leq f(D)$. Therefore we have

$$D^2/4+D \le \mathrm{f}(D) \le 2D^2+2D.$$

Question 5 [21]: Tighten the bounds for $\mathrm{f}(D)$.

In [22] Chvátal and Thomassen proved that the problem of deciding whether an undirected graph admits an orientation of diameter 2 is NP-complete. (The reader is referred to [32] for a discussion in NP-completeness.)

Recently, M. R. Garey, R. T. Tarjan and the author studied the problem of orienting all undirected edges in a mixed graph so as to minimize the diameter. It can be shown [21] that if a mixed graph of diameter D has any strongly connected orientation, then there is an orientation of diameter at most $8D^2+4D$. The proof gives a polynomial algorithm for constructing such an orientation. Suppose we define $\overline{\mathrm{f}}(d)$ similarly for the case of mixed graphs. Then we have

$$D^2/4+D \le \overline{\mathrm{f}}(D) \le 8D^2+4D$$

Question 6: Improve bounds for $\overline{\mathrm{f}}(D)$.

5. Diameter bounds for altered graphs. K. Vijayan and U. S. R. Murty [50] first investigated the following extremal problem which is motivated by constructing optimal networks with diameter and reliability considerations:

Determine the least number of edges for a graph on n vertices and diameter D having the property that, if any t edges (vertices) are deleted, the remaining graph has diameter no more than D'. Although this problem has received much attention in the past (see the surveys [3, 6, 12, 19]), it seems to be quite difficult in general, and relatively little is known beyond cases with small values of t and D (primarily $t = 1$ and $D \le 5$). We will just mention the following intriguing problem:

Question 7 [12]: How many edges must a graph have so that after removing any t edges it still has diameter $\le D$?

Motivated it by the following examples B. Bollobás [12] conjectured that such graph G must have $(1+o(1))(n + \frac{n}{\lfloor 2D/3 \rfloor})$ edges for the case $t = 1$. In particular he verified this conjecture for the case that G has diameter $< \frac{2}{3}D$, before removing edges.

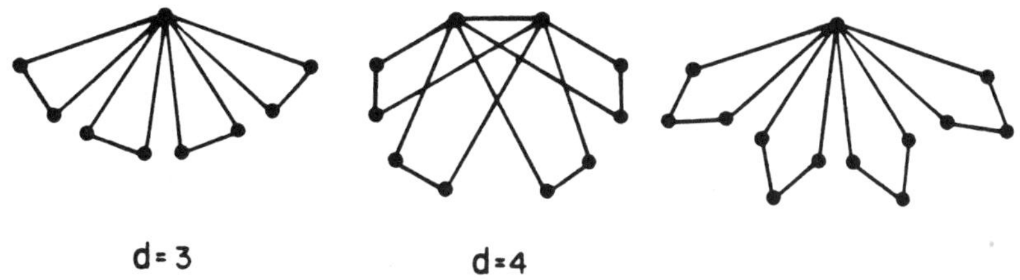

FIGURE 4

For the analogue of vertex deletion, B. Bollobás conjectured that any graph with n vertices must have at least $(1+o(1))(n + \frac{n}{\lfloor d/2 \rfloor})$ edges if it satisfies the property that after removing any edge the remaining graph has diameter $\leq D$ (see Fig. 5).

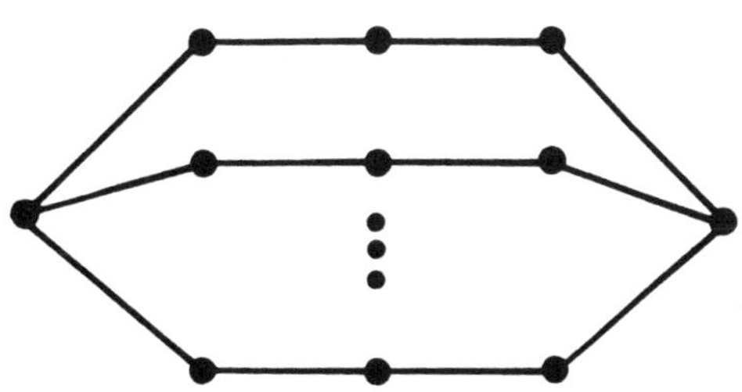

FIGURE 5

Again he verified [12] the conjecture for the case that the graph starts with diameter $< D/2$.

M. Garey and the author [20] recently studied the effect of edge or vertex deletion on the diameter bounds for general graphs. Suppose we delete an edge from a 2-connected graph, the new diameter can be twice as large as the old diameter, (deleting an edge from a cycle, for example). It is not difficult to prove that the maximum new diameter after deletion of an edge is $2D$ where D denotes the old diameter [20,42]. In general, it can be shown that when t edges are removed from a Graph G of diameter D, the resulting graph, if it is still connected, has maximum possible diameter approximately $(t+1)\cdot D$.

The corresponding vertex-deletion case is also studied in [20]. It is proved that if t vertices are deleted from a λ vertex-connected graph G with n vertices and diameter D, the resulting graph has diameter at most $(\frac{n-t-2}{\lambda-t}+1)D/2$ and this bound is best possible.

These problems are related to the following augmentation problem which is interesting in its own right. If t edges are added to a path P_n or a cycle C_n, how small can the diameter be? Let $P(n,t)$ denote this least possible diameter for the case of paths, and let $C(n,t)$ denote the corresponding value for the case of cycles, we have the following [20]:

$$\frac{n}{t+1} - 1 \le P(n,t) < \frac{n}{t+1} + 3$$

and

$$\frac{n}{t+2} - 1 \le C(n,t) < \frac{n}{t+2} + 3 \quad \text{if } t \text{ is even}$$

$$\frac{n}{t+1} - 1 \le C(n,t) < \frac{n}{t+1} + 3 \quad \text{if } t \text{ is odd}$$

Problem 8: If t edges are removed from a strongly $(t+1)$-edge-connected directed graph, how large can the diameter of the resulting graph be? This problem has been answered by C. Peyrot recently.

Problem 9: Find a fast algorithm in determining the maximum diameter of a graph after any choice of t edges are removed.

We will also mention the following interesting problem which is relevant in the sense that many extremal graphs with diameter constraints contain relatively few small cycles [46].

Problem 10 [1,18]: For given integers r and s, suppose G is a directed graph with n vertices and outdegree at least r. Is it true that if $n \le rs$ then G contains a (directed) cycle of length $\le s$?

This problem is unanswered even for the case of $s = 3$!

6. The complexity of determining the diameter.

For a given graph G on n vertices a straightforward way to determine the diameter $D(G)$ is as follows:

(1) Find the breadth-first search tree [28, 46] for each vertex v of G. Thereby determine the maximum distance $d_v = \max_u d(u,v)$.

(2) Compare d_v and determine $D(G) = \max_v d_v$.

Since the time requirement [46] for finding a breadth-first search tree is $O(n+e)$, the preceding algorithm has complexity $O(n^2+ne)$ where $e = |E(G)|$. This algorithm has in fact calculated the distances among all pairs of vertices. The problem of finding all distances is a well-studied problem in graph algorithms (see [46,47] in a somewhat general setting.) Here we will mention the fact that M. L. Fredman [30] has on $O(n^3(\log\log n/\log n)^{1/3})$ algorithm for finding distances of all pairs, which is faster than the straightforward algorithm for high density graphs.

The problem of finding the diameter of a graph is just to find the fartherest pair of vertices while pairs with short distances can be ignored. To take advantage of this, we can use the matrix multiplication algorithm to reduce the running time, which will be described in the following:

Let A denote the adjacency matrix of G, i.e., $A = (a_{ij})$ is a $n \times n$ and $a_{ij} = 1$ if and only if $[v_i, v_j]$ is an edge. (Note that G can be a mixed graph.) It is not difficult to see that in the k-th matrix product $\bar{A}^k = (A+I)^k$ of A the (i, j)-entry is nonzero if and only if there is a walk of length nor more than k from v_i to v_j. Therefore the diameter $D(G)$ is the last integer D with the property that $\bar{A}^D$ has all entries nonzero. The best known matrix multiplication algorithm due to D. Coppersmith and S. Winograd [23] requires running time $O(n^{2.496})$. The time required in computing $D(G)$ for G is then no more than $O(n^{2.496}\log D)$, since we can first find the least integer k with the property that A^{2^k} has only nonzero entries (by $n^{2.496}$ log D steps) and then use binary approximation to determine D (by another $n^{2.496}$ log D steps).

For the complexity lower bound, since every vertex and edge must be examined to determine the diameter, the obvious lower bound is $n+e$. There is of course the problem of further narrowing the gap between the upper and lower bounds on the complexity of determining $D(G)$.

Problem 11: Find a fast algorithm for determining the diameter of a graph.

Instead of finding the diameter of G, George and Lin [31] asked the question of finding a pair of pseudoperipheral vertices, i.e., a pair $\{x, y\}$ of vertices such that $d(x, y) \geq d(v,y)$ and $d(x, y) \geq d(x, v)$ for any vertex v in G

A greedy algorithm for this problem can be described as follows:

Step (1) Start from any vertex v and find a farthest vertex u from v.

Step (2) If v is also furthest from u, $\{u, v\}$ is a solution. Stop. If there is a vertex w with $d(u, w) > d(u, v)$, go to (1) and replace v by w.

One interesting question is as follows:

Problem 12: Is it true that this greedy algorithm must stop before $c\sqrt{n}$ iterations of Step (1)?

There exists a graph together with a starting point such that the greedy algorithm takes $c\sqrt{n}$ iterations (see [41]). We remark that J. K. Pachl has another algorithm for solving this problem with worse case time $\sqrt{ne}$. (Note that each iteration in the greedy algorithm takes e steps). However, the complexity of the greedy algorithm still remains open.

REFERENCES

[1] M. Behzad, G. Chartrand and C. E. Wall, On minimal regular digraphs with given girth, Fund. Math. 69 (1970), 227-231.

[2] C. Berge, Graphs and Hypergraphs. North-Holland Publishing Co, New York (1973).

[3] J. C. Bermond and B. Bollobás, The diameter of graphs - a survey, Proc. in Congressus Numerantium, vol. 32, (1981), 3-27.

[4] J. C. Bermond, C. Delorme and G. Farhi, Large graphs with given degree and diameter III. In Proc. Coll. Cambridge (1981). Ann. Discr. Math. 13, North-Holland, (1982), 23-32.

[5] J. C. Bermond, C. Delorme and J. J. Quisquater, Grands graphes de degrés et diamètre fixés. In Proc. Coll. C.N.R.S., Marseille, (1981), Ann. Discr. Math. 17, (1983), 65-73.

[6] J. C. Bermond, J. Bond. M. Paoli and C. Peyrat, Graphs and interconnection networks: diameter and vulnerability, Proc. of the Tenth British Comb. Conf. (1983), Cambridge University Press, London Math. Soc. Let. Notes Series 82 (1983), 1-30.

[7] J. C. Bermond, C. Delorme and J. J. Quisquater, Strategies for interconnection networks: Some methods from graph theory, Tech. Report of Philips Research Lab. 1984.

[8] F. Boesch and R. Tindell, Robbins theorem for mixed multigraphs. *Amer. Math. Monthly 87* (1980), 716-719.

[9] F. T. Boesch, F. Harary and J. A. Kabell, Graphs and models of communication network vulnerability: connectivity and persistence. Networks, vol. 11, (1981), 57-63.

[10] F. T. Boesch, Graph theoretic models for network reliability studies, Electrical engineering and computer science department. Technical report 8010, (1983).

[11] F. T. Boesch, Introduction to basic network problems, The Mathematics of Networks, AMS Short Course Lecture Notes (1982), 1-29.

[12] B. Bollobás, Extremal graph theory, Academic Press Inc., London (1978).

[13] B. Bollobás, The diameter of random graphs. Trans. Amer. Math. Soc. 267, (1981), 41-52.

[14] B. Bollobás and W. F. de la Vega, The diameter of random graphs. Combinatoria, 2(1982), 125-134.

[15] J. A. Bondy and U. S. R. Murty, Graph Theory with Applications. American Elsevier Publishing Co., Inc. (1976).

[16] W. G. Bridges and S. Toueg, On the impossibility of directed Moore graphs. J. of Comb. Theory, Ser B, 29, (1980), 339-341.

[17] N. G. de Bruijn, A combinatorial problem. Nederl. Akad. Wetensh. Proc. 49, (1946), 758-764.

[18] L. Cacetta and R. Häggkvist, On minimal digraphs with given girth, Proc. 8th Southeastern Conf. on Combinatorics, Graph Theory and Computing (1978), 181-187.

[19] L. Cacetta, On extremal graphs with given diameter and connectivity, In Topics in Graph Theory. Proc. New York Acad. Sc. Scientist-in-Residence program, New York 1977. Ann. New York Acad. Sci. New York, 328, (1979), 76-94.

[20] F. R. K. Chung and M. R. Garey, Diameter bounds for altered graphs. J. of Graph Theory 8 (1984), 511-534.

[21] F. R. K. Chung, M. R. Garey and R. E. Tarjan, Strongly connected orientations of mixed graphs, (to appear).

[22] V. Chvátal and C. Thomassen, Distances in orientations of graphs, J. of Comb. Theory (B), (1978), 61-75.

[23] D. Coppersmith and S. Winograd, On the asymptotic complexity of matrix multiplication, Proc. 220th Ann. Symp. on Foundation of Computer Science, Nashville (1981), 82-90.

[24] C. Delorme, Large bipartite graphs with given degree and diameter, prepoint.

[25] P. Erdös and A. Rényi, On a problem in the theory of graphs, Publ. Math. Inst. Hungar. Acad. Sci. 7, (1962), 623-641.

[26] P. Erdös, A. Rényi and V. T. Sós, On a problem of graph theory, Studia Sci. Math. Hungar. 1, (1966), 215-235.

[27] P. Erdös, S. Fajtlowicz and A. J. Hoffman, Maximum degree in graphs of diameter 2. Networks, vol. 10, (1980), 87-90.

[28] S. Even, *Graph Algorithms*, Computer Science Press, Potomac, Md. (1979).

[29] W. Feit and G. Higman, The non-existence of certain generalized polygons, J. Algebra 1, (1964), 114-131.

[30] M. L. Fredman, New bounds on the complexity of the shortest path problem, SIAM J. Comput., 5 (1976), 83-89.

[31] A. George and J. W. H. Lin, Computer Solution of Large Sparse Positive Definite Systems, Prentice-Hall, Englewood Cliffs, NJ 1981.

[32] M. Garey and D. Johnson, *Computers and Intractabiality*, Freeman, San Francisco, CA (1979).

[33] F. Harary, *Graph Theory*, Addison-Wesley, Reading, Mass. (1969,

[34] A. J. Hoffman and R. R. Singleton, On Moore graphs with diameter 2 and 3, IBM J. of Res. Development 4 (1960), 497-504.

[35] M. Jerrum and S. Skyum, Families of fixed degree graphs for processor interconnection, IEEE Trans. on Computers 33 (1984), 190-194.

[36] W. Kantor, Generalized polygons, SCABS and GABS, preprint (to appear in Springer Lecture Notes, Proc. of a geometry conference in Italy).

[37] G. Katona and E. Szemerédi, On a problem of graph theory, Studia Sci. Math. Hungar. 2 (1967), 23-28.

[38] W. Leland, R. Finkel, L. Qiao, M. Solomon and L. Uhr, High density graphs for processor interconnection. Inf. Proc. Letters, vol. 12, (1981), 117-120.

[39] W. Leland and M. Solomon, (1982). Dense trivalent graphs for processor interconnection. IEEE Trans. Comp. 31, no. 3, 219-222.

[40] J. Pach and L. Surányi, Graphs of diameter 2 and linear programming. In Proc. Colloq. Algebraic Methods in Graph Theory, (eds. L. Lovász and V. T. Sós), North-Holland, Coll. Math. Soc. Janos Bolyai, 25 Vol. 2 (1981), 599-629.

[41] J. K. Pachl, Finding Pseudoperipheral nodes in graphs, J of Computer and System Sciences 29, (1984), 48-53.

[42] J. Plesnik, Note on diametrically critical graphs. In Recent Advances in Graph Theory, Proc. Symp. Prague, Academia Prague, (1975), 455-465.

[43] J. Plesnik and S. Znam, Strongly geodetic directed graphs, Acta Fac Rerum Natur. Univ. Camenian Math. 29, (1974) 29-34.

[44] S. M. Reddy, J. G. Kuhl, S. H. Hosseini and H. Lee, On diagraph with minimum diameter and maximum connectivity, Proceedings of Allerton Conf. on communications, control and computers, Illinois (1982).

[45] H. Robbins, A theorem on graphs with an application to a problem of traffic control, *Amer. Math. Monthly 46* (1939), 281-283.

[46] R. E. Tarjan, Minimum spanning trees, unpublished memorandum, Bell Laboratories (1981).

[47] R. E. Tarjan, Data Structures and Network Algorithms, SIAM Publications 1983.

[48] J. Tits, "Suc la trialité et certains groupes qui Sé déduisent, Inst. Hautes Etudes Sci. Publ. Math. 2 (1959), 14-60.

[49] J. A. Thas, Polar spaces, generalized hexagons, and perfect codes, J.C.T. (A) 29 (1980), 87-93.

[50] K. Vijayan and U. S. R. Murty, On accessibility in graphs, Sankhya Ser. A. 26 (1964), 299-302.

[51] P. K. Wong, Cages - A survey, J. of Graph Theory 6, (1982), 1-22.

[52] A. Yanushka, On order in generalized polygons, Geometrical Dedicata 20 (1981), 451-458.

Proceedings of Symposia in Applied Mathematics
Volume 34, 1986

THE THEORY OF DATA DEPENDENCIES - A SURVEY

Ronald Fagin and Moshe Y. Vardi

Abstract: The language of database dependencies can be seen as a language for specifying the semantics of databases. Dependency theory studies the properies of this language and its use in database management systems. We survey here three aspects of dependency theory: the *implication problem*, the *universal relation model*, and *acyclicity* of database schemes.

1. Introduction

In the *relational database model*, conceived by Codd in the late 60's [Co1], one views the database as a collection of relations, where each relation is a set of tuples over some domain of values. One notable feature of this model is its being almost devoid of semantics. A tuple in a relation represents a relationship between certain values, but from the mere syntactic definition of the relation one knows nothing about the nature of this relationship, not even if it is a one-to-one or one-to-many relationship.

One approach to remedy this deficiency is to devise means to specify the missing semantics. These semantic specifications are often called *semantic* or *integrity constraints*, since they specify which databases are meaningful for the application and which are meaningless. Of particular interest are the constraints called *data dependencies*, or dependencies for short.

1980 *Mathematical Subject Classification*. Primary 68B15, Secondary 03B25.

The study of dependencies began in 1972 with the introduction by Codd [Co2] of the *functional dependencies*. After the introduction, independently by Fagin and Zaniolo [Fa1,Za] in 1976, of *multivalued dependencies*, the field became chaotic for a few years in which researchers introduced many new classes of dependencies. The situation has stabilized since 1980 with the introduction, again independently by various researchers, of *embedded implicational dependencies* (EIDs). Essentially, EIDs are sentences in first-order logic stating that if some tuples, fulfilling certain equalities, exist in the database then either some other tuples must also exist in the database or some values in the given tuples must be equal. The class of EIDs seems to contain most previously studied classes of dependencies. (Recently, De Bra and Paredaens [DP] considered *afunctional dependencies*, which are not EIDs.) We give basic definitions and historical perspective in Section 2.

Most of the papers in dependency theory deal exclusively with various aspects of the *implication problem*, i.e., the problem of deciding for a given set of dependencies Σ and a dependency τ whether Σ *logically implies* τ. The reason for the prominence of this problem is that an algorithm for testing implication of dependencies enables us to test whether two given sets of dependencies are equivalent or whether a given set of dependencies is redundant. A solution for the last two problems seems a significant step towards automated database schema design, which some researchers see as the ultimate goal for research in dependency theory [BBG]. We deal with the implication problem in Section 3.

An emerging application for the theory of dependencies is the *universal relation model*. This model aims at achieving *data independence*, which was the original motivation for the relational model. In the universal relation model the user views the data as if it is stored in one big relation. The data, however, is not available in this form but rather in several smaller relations. It is the role of the database management system to provide the interface between the users' view and the actual data, and it is the role of the database designer to specify this interface. There have been different approaches to the question of what this interface should be like. We describe one approach, the *weak* universal relation approach, in Section 4.

One of the the major contributions of theoretical computer science is delineate the line between the computationally feasible and the infeasible.

The plethora of unsolvability and intractability results forces researchers to lower their aims, and to restrict themselves to "real-world" problems. That is, rather than trying to solve problems in their full generality, they try solve the cases that most often arise in practice. Research on *acyclicity* of database schemes falls in this vein. The idea is to view database schemes as *hypergraphs.* It turns out the acyclicity properties for hypergraphs translate into highly desirable database properties. Many problems that are intractable for general database schemes can be solved quite efficiently for acyclic schemes. Furthermore, there are arguments that many applications can be represented by acyclic schemes. We discuss acyclicity in Section 5.

A survey like ours of a rich theory necessarily has to be selective. The selection naturally reflects our tastes and biases. A more comprehensive, though less up to date, coverage can be found in the books [Ma,Ul].

2. Definitions and historical perspective

We begin with some fundamental definitions about relations. We are given a fixed finite set U of distinct symbols, called *attributes,* which are column names. From now on, whenever we speak of a set of attributes, we mean a subset of U. Let R be a set of attributes. An *R-tuple* (or simply a *tuple,* if R is understood) is a function with domain R. Thus, a tuple is a mapping that associates a value with each attribute in R. Note that under this definition, the "order of the columns" does not matter. If S is a subset of R, and if t is an R-tuple, then $t[S]$ denotes the S-tuple obtained by restricting the mapping to S. An *R-relation* (or a *relation over R*, or simply a *relation,* if R is understood), is a set of R-tuples. In database theory, we are most interested in *finite* relations, which are finite sets of tuples (although it is sometimes convenient to consider infinite relations). If I is an R-relation, and if S is a subset of R, then by $I[S]$, the *projection* of I onto S, we mean the set of all tuples $t[S]$, where t is in I. A *database* is a finite collection of relations.

Conventions: Upper-case letters A,B,C, ... from the start of the alphabet represent single attributes; upper-case letters $R,S,...,Z$ from the end of the alphabet represent sets of attributes; upper-case letters $I,J,...$ from the middle of the alphabet represent relations; and lower-case letters $r,s,t,...$ from the end of the alphabet represent tuples.

Assume that relations $I_1,...,I_n$ are over attribute sets $R_1,...,R_n$ respectively. The *join* of the relations $I_1,...,I_n$, which is written either $\bowtie \{I_1,...,I_n\}$ or $I_1 \bowtie ... \bowtie I_n$, is the set of all tuples t over the attribute set $R_1 \cup ... R_n$, such that $t[R_i]$ is in I_i for each i. (Our notation exploits the fact that the join is associative and commutative.)

Certain sentences about relations are of special practical and/or theoretical interest for relational databases. For historical reasons, such sentences are usually called *dependencies.* The first dependency introduced and studied was the *functional dependency* (or FD), due to Codd [Co2]. As an example, consider the relation in Figure 2.1, with three columns: EMP (which represents employees), DEPT (which represents departments), and MGR (which represents managers). The relation in Figure 2.1 obeys the FD "DEPT→MGR", which is read "DEPT determines MGR". This means that whenever two tuples (that is, rows) agree in the DEPT column, then they necessarily agree also in the MGR column. The relation in Figure 2.2 does not obey this FD, since, for example, the first and fourth tuples agree in the DEPT column but not in the MGR column. We now give the formal definition. Let X and Y be subsets of the set U of attributes. The FD $X \rightarrow Y$ is said to hold for a relation I if every pair of tuples of I that agree on each of the attributes in X also agree in the attributes in Y.

The original motivation for introducing FDs (and some of the other dependencies we discuss) was to describe database *normalization.* Before giving an example of normalization, we need to define the notion of a relation scheme. A *relation scheme* is simply a set R of attributes. Usually, there is also an associated set Σ of sentences about relations over R. A relation is an *instance* of the relation scheme if it is over R and obeys the sentences in Σ. Thus, the sentences Σ can be thought of as "constraints", that every "valid instance" must obey. Although we do not do so, we note that it is common to define a relation scheme to be a pair $<R,\Sigma>$, where the constraints Σ are explicitly included.

We now consider an example of normalization. Assume that the attributes are {EMP,DEPT,MGR}, and that the only constraint is the FD DEPT→MGR. So, in every instance of this scheme, two employees in the same department necessarily have the same manager. It might be better to store the data not in one relation, as in Figure 2.1, but rather in two relations, as in Figure 2.3: one relation that relates employees to departments, and one relation that relates departments to managers. We shall come back to normalization in Section 4.

It is easy to see that FDs can be represented as sentences in first-order logic [Ni1]. Assume, for example, that we are dealing with a 4-ary relation, where the first, second, third, and fourth columns are called, respectively, A, B, C, and D. Then the FD $AB \rightarrow C$ is represented by the following sentence:

$$(\forall abc_1c_2d_1d_2)((Pabc_1d_1 \wedge Pabc_2d_2) \Rightarrow (c_1 = c_2)). \qquad (2.1)$$

Here $(\forall abc_1c_2d_1d_2)$ is shorthand for $\forall a \forall b \forall c_1 \forall c_2 \forall d_1 \forall d_2$, that is, each variable is universally quantified. Unlike Nicolas, we have used individual variables rather than tuple variables. Incidentally, we think of P in (2.1) as a *relation symbol*, which should not be confused with an *instance* (that is, a *relation*) I, for which (2.1) can hold.

Let X and Y be sets of attributes (subsets of U), and let Z be U-XY (by XY, we mean $X \cup Y$). Thus, Z is the set of attributes not in X or Y. As we saw by example above (where X, Y, and Z are, respectively, the singleton sets {DEPT}, {EMP}, and {MGR}), the FD $X \rightarrow Y$ is a sufficient condition for a "lossless decomposition" of a relation with attributes U into two relations, with attributes XY and XZ respectively. This means that if I is a relation with attributes XYZ that obeys the FD $X \rightarrow Y$, then I can be obtained from its projections $I[XY]$ and $I[XZ]$, by joining them together. Thus, there is no loss of information in replacing relation I by the two relations I_1 and I_2. We note that this fact, which is known as Heath's Theorem [He], is historically one of the first theorems of database theory.

It may be instructive to give an example of a decomposition that does lose information. Let I be the relation in Figure 2.4, with attributes STORE, ITEM, and PRICE. Let I_1 and I_2 be two projections of I, onto {STORE, ITEM} and {ITEM, PRICE}, respectively, as in Figure 2.5. These projections contain less information than the original relation I. Thus, we see from relation I_1 that Macy's sells toasters; further, we see from relation I_2 that someone sells toasters for 20 dollars, and that someone sells toasters for 15 dollars. However, there is no way to tell from relations I_1 and I_2 how much Macy's sells toasters for.

The next dependency to be introduced was the *multivalued dependency*, or MVD, which was defined, independently by Fagin [Fa1] and Zaniolo [Za]. It was introduced because of the perception that the functional dependency provided too limited a notion of "depends on". As we shall see, multivalued dependencies provide a necessary and sufficient condition for lossless decomposition of a relation into two of its projections. Before we give the formal definition, we present a few examples. Consider the relation in Figure 2.6, with attributes EMP, SALARY, and CHILD. It obeys the functional dependency EMP→SALARY, that is, each employee has exactly one salary. The relation does *not* obey the FD EMP→CHILD, since an employee can have more than one child. However, it is clear that in some sense an employee "determines" his set of children. Thus, the employee's set of children is "determined by" the employee and by nothing else, just as his salary is. Indeed, as we shall see, the multivalued dependency EMP↠CHILD (read "employee multidetermines child") holds for this relation. As another example, consider the relation in Figure 2.7, with attributes EMP, CHILD, and SKILL. A tuple (e,c,s) appears in this relation if and only if e is an employee, c is one of e's children, and s is one of e's skills. This relation obeys no nontrivial (nontautologous) functional dependencies. However, it turns out to obey the multivalued dependencies EMP↠CHILD and EMP↠SKILL. Intuitively, the MVD EMP↠CHILD means that the set of names of the employee's children depends only on the employee, and is "orthogonal" to the information about his skills.

We are now ready to formally define multivalued dependencies. Let I be a relation over U. As before, let X and Y be subsets of U, and let Z be U-XY. The multivalued dependency $X \twoheadrightarrow Y$ holds for relation I if for each pair r, s of tuples of I for which $r[X] = s[X]$, there is a tuple t in I where (1) $t[X] = r[X] = s[X]$, (2) $t[Y] = r[Y]$, and (3) $t[Z] = s[Z]$. Of course, if this multivalued dependency holds for I, then it follows by symmetry that there is also a tuple u in I where (1) $u[X] = r[X] = s[X]$, (2) $u[Y] = s[Y]$, and (3) $u[Z] = r[Z]$.

Multivalued dependencies obey a number of useful properties. For example, if U is the disjoint union of X, Y, Z, and W, and if I is a relation over U that obeys the MVDs $X \twoheadrightarrow Y$ and $Y \twoheadrightarrow Z$, then it follows that I obeys the MVD $X \twoheadrightarrow Z$ [Fa1]. So, MVDs obey a law of transitivity. We shall discuss more properties of MVDs in Section 3, where we give a complete axiomatization for MVDs.

Note that MVDs, like FDs, can be expressed in first-order logic. For example, assume that $U=\{A,B,C,D,E\}$. Then the MVD $AB \twoheadrightarrow CD$ holds for a relation over U if the following sentence holds, where P plays the role of the relation symbol:

$$(\forall abc_1c_2d_1d_2e_1e_2)\ ((Pabc_1d_1e_1 \wedge Pabc_2d_2e_2) \Rightarrow Pabc_2d_2e_1). \tag{2.2}$$

Embedded dependencies were introduced (Fagin [Fa1]) as dependencies that hold in a projection of a relation (although, as we shall see, for certain classes of dependencies they are defined a little more generally). We shall simply give an example of an embedded MVD; the general case is obvious from the example. Assume that we are dealing with 4-ary relations, where we call the four columns $ABCD$. We say that such a 4-ary relation I obeys the *embedded MVD* (or *EMVD*) $A \twoheadrightarrow B \mid C$ if the projection of R onto ABC obeys the MVD $A \twoheadrightarrow B$. Thus, the EMVD $A \twoheadrightarrow B \mid C$ can be written as follows:

$$(\forall ab_1b_2c_1c_2d_1d_2)((Pab_1c_1d_1 \wedge Pab_2c_2d_2) \Rightarrow \exists d_3 Pab_1c_2d_3). \tag{2.3}$$

As a concrete example, assume that the relation of Figure 2.7, with attributes EMP, CHILD, and SKILL, had an additional attribute BIRTHDATE, which tells the date of birth of the child. Then this 4-ary relation I would obey the embedded MVD EMP$\twoheadrightarrow$CHILD | SKILL. Note that I need not obey the MVD EMP$\twoheadrightarrow$CHILD (although it does obey the MVD EMP$\twoheadrightarrow$ {CHILD,BIRTHDATE}).

Several dependencies were defined within a few years after the multivalued dependency was introduced; we shall mention these other dependencies later in this section. Of these, the most important are the *join dependency*, or JD [ABU,Ri2]), and the *inclusion dependency*, or IND [Fa2]. Assume that $\mathbf{X}=\{X_1, \ldots, X_k\}$ is a collection of subsets of U, where $X_1 \cup \ldots \cup X_k = U$. The relation I, over U, is said to obey the join dependency $\bowtie [X_1, \ldots, X_k]$, denoted also $\bowtie [\mathbf{X}]$, if I is the join of its projections $I[X_1],\ldots,I[X_k]$. It follows that this join dependency holds for the relation I if and only if I

contains each tuple t for which there are tuples $w_1, ..., w_n$ of I (not necessarily distinct) such that $w_i[X_i] = t[X_i]$ for each i $(1 \leq i \leq n)$. As an example, consider the relation I in Figure 2.8 below.

A	*B*	*C*	*D*
0	1	0	0
0	2	3	4
5	1	3	0

Figure 2.8

This relation violates the join dependency $\bowtie [AB, ACD, BC]$. For, let w_1, w_2, w_3 be, respectively, the tuples (0,1,0,0), (0,2,3,4), and (5,1,3,0) of I; let X_1, X_2, X_3 be, respectively, AB, ACD, and BC; and let t be the tuple (0,1,3,4); then $w_i[X_i] = t[X_i]$ for each i $(1 \leq i \leq n)$, although t is not a tuple in the relation I. However, it is straightforward to verify that the same relation I obeys, for example, the join dependency $\bowtie [ABC, BCD, ABD]$.

Let us say that the join dependency $\bowtie [X_1, ..., X_k]$ has k *components.* Join dependencies are generalizations of multivalued dependencies; thus, each multivalued dependency is equivalent to a join dependency with two components, and conversely. Assume now that $X_1 \cup ... \cup X_k \subseteq U$, and denote $X_1 \cup ... \cup X_k$ by X. A relation I with attributes U is said to obey the *embedded join dependency* $\bowtie [X_1, ..., X_k]$ if its projection $I[X]$ obeys the join dependency $\bowtie [X_1, ..., X_k]$. We shall see soon that join dependencies can be written in first-order logic. Embedded join dependencies, too, can be so written, but they require existential quantifiers, just as embedded multivalued dependencies do. Note that our notation, the set U of attributes does not appear, and so the same syntactical object $\bowtie [X_1, ..., X_k]$ is used to represent a join dependency over X and an embedded join dependency over U. However, the two would be written in distinct ways in first-order logic. This is actually a nice convenience, especially in the case of functional dependencies, where a similar comment applies.

The intuitive semantics of *multivalued* dependencies were fairly well understood at the time they were first defined. However, it was not until several years after *join* dependencies were defined that their semantics was adequately explained (by Fagin et al. [FMU]). Let us consider an example (from [FMU]). Assume that the attributes are C(ourse), T(eacher), R(oom), H(our), S(tudent), and G(rade). The informal meaning of these attributes is that teacher T teaches course C, course C meets in room R at hour H, and that student S is getting grade G in course C. If we were to define a single "universal" relation over these attributes, it would be

$\{(c,t,r,h,s,g)$: t "teaches" c; c "meets in" r "at hour" h; and s " is getting" g "in" $c\}$.

This relation is of the form

$$\{(c,t,r,h,s,g): P_1tc \wedge P_2crh \wedge P_3sgc\}, \tag{2.4}$$

for certain predicates P_1, P_2, and P_3. The fact that a relation with attributes c,t,r,h,s,g is of the form (2.4) for for some predicates P_1, P_2, and P_3 is a severe constraint. In fact [FMU], this constraint is precisely equivalent to the join dependency $\bowtie[TC,CRH,SGC]$. The obvious generalization of this observation to arbitrary join dependencies explains their semantics. Before we leave join dependencies, let us note, as promised, they, too, can be written as sentences in first-order logic. For example, if we are dealing with relations with attributes c,t,r,h,s,g, then the join dependency $\bowtie[TC,CRH,SGC]$ can be written as

$$(\forall ctt_1t_2rr_1r_2hh_1h_2ss_1s_2gg_1g_2)$$
$$((Pctr_1h_1s_1g_1 \wedge Pct_1rhs_2g_2 \wedge Pct_2r_2h_2sg) \Rightarrow Pctrhsg) \tag{2.5}$$

So far, each of the dependencies we have discussed has two properties: (1) each is *uni-relational*, that is, deals with a single relation at a time, rather than with inter-relationships among several relations, and (2) each is typed. By *typed*, we mean that no variable appears in two distinct columns. For example, the sentence $(\forall xy)((Pxy \wedge Pyz) \Rightarrow Pxz)$, which says that a relation is transitive, is *not* typed, since the variable y appear in both the first and second columns of P in the sentence. The next dependency that we shall discuss violates both (1) and (2) above, that is, is neither uni-relational nor typed. This dependency is the *inclusion dependency*, or IND [CFP]. As an example, an IND can say that every MANAGER entry of the P relation appears as an EMPLOYEE entry of the Q relation. In general, an IND is of the form

$$P[A_1..A_m] \subseteq Q[B_1...B_m], \tag{2.6}$$

where P and Q are relation names (possibly the same), and where the A_i's and B_i's are attributes. If I is the P relation and J is the Q relation, then the IND (2.6) holds if for each tuple s of I, there is a tuple t of J such that $s[A_1..A_m] = t[B_1...B_m]$. Hence, INDs are valuable for database design, since they permit us to selectively define what data must be duplicated in what relations. INDs are commonly known in Artificial Intelligence applications as *ISA* relationships (cf. Beeri and Korth [BK]). Not surprisingly, the inclusion dependency, too, can be written in first-order logic. For example, if the P relation has attributes ABC, and the Q relation has attributes CDE, then the IND $P[AB] \subseteq Q[CE]$ can be written

$$(\forall abc)(Pabc \Rightarrow \exists dQadb). \tag{2.7}$$

After multivalued dependencies were defined, there was a period where a large number of other dependencies were defined. We have already discussed the classes of join dependencies, embedded join dependencies, and inclusion dependencies. Others (many of which were introduced before join dependencies) include Nicolas's *mutual dependencies* [Ni1], which say that a relation is the join of three of its projections, Mendelzon and Maier's *generalized mutual dependencies* [MM], Paredaens' *transitive dependencies* [Pa], which generalize both FDs and MVDs, Ginsburg and Zaiddan's *implied dependencies* [GZ], which generalize FDs, Sagiv and Walecka's *subset dependencies* [SW], which generalize embedded MVDs, Sadri and Ullman's and Beeri and Vardi's *template dependencies* ([SU], [BV4]) which generalize embedded join dependencies, and Parker and Parsaye-Ghomi's *extended transitive dependencies* [PP], which generalize both mutual dependencies and transitive dependencies. We remark that the last 3 kinds of dependencies mentioned were introduced to deal with the issue of a complete axiomatization (see Section 3): subset dependencies were introduced to show the difficulty of completely axiomatizing embedded multivalued dependencies; extended transitive dependencies were introduced to show the difficulty of completely axiomatizing transitive dependencies; while template dependencies were introduced to provide a class of dependencies that include join dependencies and that can be completely axiomatized. Inclusion dependencies, which had been used informally for databases by many practitioners, were not seriously studied until relatively late [CFP].

Various researchers finally realized that all of these different types of dependencies can be united into a single class, which we shall call simply *dependencies*. Before we can define them formally, we need a few preliminary concepts. We assume that we are given a set of *individual variables* (which represent entries in a relation of a database). The *atomic* formulas are those that are either of the form $Pz_1...z_d$ (where P is the name of a d-ary relation, and where the z_i's are individual variables), or else of the form $x=y$ (where x and y are individual variables). Atomic formulas $Pz_1...z_d$ we call *relational formulas*, and atomic formulas $x=y$ we call *equalities*. A *dependency* is a first-order sentence

$$(\forall x_1...x_m)((A_1 \wedge ... \wedge A_n) \Rightarrow \exists y_1...y_r(B_1 \wedge ... \wedge B_s)), \tag{2.8}$$

where each A_i is a relational formula and where each B_i is atomic (either a relational formula or an equality). We assume also that each of the x_j's appears in at least one of the A_i's, and that $n \geq 1$, that is, that there is at least one A_i. We assume that $r \geq 0$ (if $r=0$ then there are no existential quantifiers), and that $s \geq 1$ (that is, there must be at least one B_i.) Note that because of all these assumptions, each dependency is obeyed by an empty database with no tuples. Furthermore, our assumptions guarantee that we can tell if a dependency holds for a relation by simply considering the collection of tuples of the relation, and ignoring any underlying "domains of attributes". Intuitively, in considering whether a dependency holds for a relation, the quantifiers can be assumed to range over the elements that appear in the relation, and not over any larger domain. This property is called *domain independence*. See Fagin [Fa4] for a much more complete discussion of domain independence.

If each of the formulas B_i on the right-hand side of (2.8) is a relational formula, then we call the dependency a *tuple-generating dependency*; if all of these formulas are equalities, then we call the dependency an *equality-generating dependency*. Of the dependencies we have focused on above, the (embedded) multivalued dependency, the (embedded) join dependency, and the inclusion dependency are each tuple-generating dependencies; thus, each of the first-order sentences (2.2), (2.3), (2.5), and (2.7) above represent tuple-generating dependencies. Tuple-generating dependencies say that if a certain pattern of entries appears, then another pattern must appear. Functional dependencies, as we see by example in the sentence (2.1) above, are equality-generating dependencies. Equality-generating dependencies say that if a certain pattern of entries appears, then a certain equality must hold. A *full* dependency is one in which r=0 and s=1 in (2.8), that is, one in which there are no existential quantifiers and in which there is only one atomic formula B_i on the right-hand side. Thus, a full dependency is of the form

$$(\forall x_1 \ldots x_m)((A_1 \wedge \ldots \wedge A_n) \Rightarrow B), \qquad (2.9)$$

where each A_i is a relational formula, where B is atomic. Functional, multivalued, and join dependencies are all full dependencies. We may refer to a dependency (2.8) as an *embedded dependency*, to emphasize that we are allowing (but not requiring) existential quantifiers. Note that in the case of full dependencies, we would not gain anything by allowing the possibility of having several atomic formulas on the right-hand side, since such a sentence is equivalent to a finite set of full dependencies as we have defined them.

The class of dependencies was defined independently by a number of authors, who usually focussed on the uni-relational case. (Note that the only special case of a dependency that we have mentioned so far that is not uni-relational is the inclusion dependency.) Beeri and Vardi [BV7] refer to this class as the class of all *tuple-generating* and *equality-generating dependencies*. Fagin [Fa4] focused on the typed, uni-relational case, which he called *embedded implicational dependencies* (with the full dependencies being called *implicational dependencies*). Yannakakis and Papadimitriou [YP] defined *algebraic dependencies*, which are built out of expressions involving projection and join, and which, on the surface, look very different from our first-order definition. It is somewhat surprising that their class (which is typed) turns out [YP] to be identical to our typed, uni-relational dependencies. Paredaens and Janssens [PJ] defined *general dependencies*, which are full, typed, uni-relational dependencies. Also, Grant and Jacobs [GJ] defined *generalized dependency constraints,* which are full dependencies.

An often heard claim is that in the "real world" one rarely encounters dependencies in their most general form. According to this claim FDs, INDs, maybe MVDs are the only kinds of dependencies that earn the title "real world dependencies". We have two answers to this claim. First, we believe that there are real-world situations that do require the more general dependencies. Even when the database itself can be specified by FDs, user views of this database may not be specifiable by FDs [Fa4, GZ]. Furthermore, even if only simple dependencies arise in practice, the more general dependencies are very useful theoretically. For example, statements about equivalence of queries can be expressed by dependencies [YP]. We refer the reader to [Hu, Va3, Va5] for more examples of the latter argument.

3. The Implication Problem

3.1. Implication and finite implication

Logical implication is a fundamental notion in logic. Let Σ be a set of sentences, and let τ be a single sentence. We say that Σ *implies* τ, denoted $\Sigma \models \tau$, if every model of Σ is also a model of τ. In our context, $\Sigma \models \tau$ if every database that satisfies all dependencies in Σ satisfies also τ. For example $\{A \rightarrow B, B \rightarrow C\} \models A \rightarrow C$.

The relevance of implication to database theory became apparent in Bernstein's work on synthesis of database schemes using FDs [Ber]. Let Σ_1 and Σ_2 be sets of dependencies. We say that Σ_1 is *equivalent to* Σ_2, denoted $\Sigma_1 \equiv \Sigma_2$, if every database that satisfies all dependencies in Σ_1 also satisfies all dependencies in Σ_2 and vice versa. We say that Σ_1 is *redundant* if $\Sigma_2 \subset \Sigma_1$ and $\Sigma_1 \equiv \Sigma_2$. (We use $\subseteq$ to denote containment and $\subset$ to denote proper containment.) Clearly, Σ is redundant if there is some $\tau \in \Sigma$ such that $\Sigma - \{\tau\} \models \tau$. Since Bernstein's synthesis algorithm requires eliminating redundant FDs , and since the problem of eliminating redundant dependencies can be reduced to the problem of testing implication of dependencies, the notion of implication became a central notion to dependency theory. The significance of implication was reconfirmed in later works, e.g., [BMSU,Ri2].

In database theory we often like to restrict our attention to finite databases, since in practice databases are finite. We say that Σ *finitely implies* τ, denoted $\Sigma \models_f \tau$, if every finite database that satisfies Σ satisfies also τ. Clearly, if $\Sigma \models \tau$ holds then $\Sigma \models_f \tau$ also holds. But it is possible that $\Sigma \models_f \tau$ holds while $\Sigma \models \tau$ does not. That is, it is possible that every finite database that satisfies Σ satisfies also τ, but there is an *infinite* database that satisfies Σ but not τ. Implication and finite implication lead to two decision problems. The *implication problem* is to decide, for a given set Σ of dependencies and a single dependency τ, whether $\Sigma \models \tau$. The *finite implication problem* is to decide, for a given set Σ of dependencies and a single dependency τ, whether $\Sigma \models_f \tau$.

Let $\Sigma = \{\sigma_1,...,\sigma_n\}$. Then $\Sigma \models \tau$ ($\Sigma \models_f \tau$) if and only if $\sigma_1 \wedge ... \wedge \sigma_n \wedge \neg \tau$ is (finitely) unsatisfiable. (A sentence is (finitely) satisfiable if it has a (finite) model. It is (finitely) unsatisfiable if it has no (finite) model.) Since unsatisfiability is known to be recursively enumerable (Gödel's Completeness Theorem), and finite satisfiability is clearly recursively enumerable, it follows that the relationships $\models$ and $\not\models_f$ are recursively enumerable. Suppose now that for some class of dependencies $\models$ and $\models_f$ are the same. Then $\models$ and $\not\models_f$ complement each other and they are both recursively enumerable. It follows that they are both recursive [Ro]. Indeed, the standard technique for proving solvability of the implication problem is to show that implication and finite implication coincide.

Dependencies are $\forall^*\exists^*$ sentences, i.e., they are equivalent to sentences whose quantifier prefix consists of a string of universal quantifiers followed by a string of existential quantifiers. Thus, $\sigma_1 \wedge ... \wedge \sigma_n \wedge \neg \tau$ is a $\exists^*\forall^*\exists^*$ sentence. When Σ, however, consists of full dependencies, then $\sigma_1 \wedge ... \wedge \sigma_n \wedge \neg \tau$ is an $\exists^*\forall^*$ sentence. Thus, the (finite) implication problem for full dependencies is reducible to the (finite) satisfiability problem for $\exists^*\forall^*$ sentences. This class of sentences is known as the *initially extended Bernays-Schönfinkel class.* For this class, satisfiability and finite satisfiability coincide, and therefore both are recursive [DG]. Thus, for full dependencies, implication and finite implication coincide, and are recursive. Unfortunately, the satisfiability problem for the Bernays-Schönfinkel class require nondeterministic exponential time [Le], and hence is highly intractable. Since the class of full dependencies is a proper subset of the class of universal sentences, one may hope that the implication problem for full dependencies is not that hard. We study this problem in Section 3.2.

For simplicity we restrict ourselves in the sequel to uni-relational dependencies, i.e., dependencies that refer to a single relation.

3.2. The implication problem for full dependencies

Since for full dependencies implication and finite implication coincide, everything we say in this section about implication holds, of course, for finite implication as well.

Even though the significance of implication was not yet clear in 1974, it was studied by Armstrong [Arm], apparently just out of mathematical interest. Armstrong characterized implication of FDs using an *axiom system.* An axiom system consists of *axiom schemes* and *inference rules.* A *derivation* of a dependency τ from a set Σ of dependencies is a sequence $\tau_1, \tau_2, \ldots, \tau_n$, where τ_n is τ and each τ_i is either an instance of an axiom scheme or follows from preceding dependencies in the sequence by one of the inference rules. $\Sigma \vdash \tau$ denotes that there is a derivation of τ from Σ. An axiom system is *sound* if $\Sigma \vdash \tau$ entails $\Sigma \models \tau$, and it is *complete* if $\Sigma \models \tau$ entails $\Sigma \vdash \tau$. Armstrong's system, denoted $\mathcal{FD}$, consists of one axiom and three inference rules:

FD0 (reflexivity axiom): $\vdash X \rightarrow X$.
FD1 (transitivity): $X \rightarrow Y,\ Y \rightarrow Z \ \vdash\ X \rightarrow Z$.
FD2 (augmentation and projection): $X \rightarrow Y \ \vdash\ W \rightarrow Z$, if $X \subseteq W$ and $Y \supseteq Z$.
FD3 (union): $X \rightarrow Y,\ Z \rightarrow W \ \vdash\ XZ \rightarrow YW$.

Theorem 3.2.1. [Arm] The system $\mathcal{FD}$ is sound and complete for implication of FDs. □
(In fact, Armstrong proved a somewhat stronger result, which we shall not discuss here. See [Fa3].)

Armstrong did not consider the algorithmic aspects of his axiom system. This was done by Beeri and Bernstein [BB], who were motivated by the fact that one of the steps in Bernstein's synthesis algorithm [Ber] is a test for implication. They were the first to phrase the implication problem. (Beeri and Bernstein called it the *membership problem.* In some papers it is also called the *inference problem.)*

Let Σ be a set of dependencies, and let X be a set of attributes. The *closure* of X with respect to Σ is the set of all attributes functionally determined by X, that is, $\{A: \Sigma \models X \rightarrow A\}$. Clearly, once we know the closure of X with respect to Σ, we can find out easily whether $\Sigma \models X \rightarrow A$. Beeri and Bernstein showed that the system $\mathcal{FD}$ can be used to construct closures very fast.

Theorem 3.2.2. [BB] The implication problem for FDs can be solved in time $O(n)$, where n is the length of the input. □

A large part of dependency theory since 1976 was devoted to studying these two aspects of implication, i.e., axiomatization and complexity of the implication problem. For example, shortly after the introduction of MVDs in 1976, they were axiomatized by Beeri et al. [BFH], and Beeri proved that implication problem is solvable [Bee]. Both works tried to get results analogous to the results for FDs.

The axiom system $\mathcal{MVD}$ consists of one axiom and three inference rules:

MVD0 (reflexivity axiom): $\vdash X \twoheadrightarrow Y$, if $Y \subseteq X$.
MVD1 (transitivity): $X \twoheadrightarrow Y,\ Y \twoheadrightarrow Z \vdash X \twoheadrightarrow Z\text{-}Y$.
MVD2 (augmentation): $X \twoheadrightarrow Y \vdash XW \twoheadrightarrow YZ$ if $Z \subseteq W$.
MVD3 (complementation): $X \twoheadrightarrow Y \vdash X \twoheadrightarrow Z$, if $XYZ = U$ and $Y \cap Z \subseteq X$.

Theorem 3.2.3. [BFH] The system $\mathcal{MVD}$ is sound and complete for implication of MVDs. □

We note that Beeri et al. [BFH] also present a sound and complete axiomatization for FDs and MVDs taken together. This axiomatization contains all of the axiom schemes and inference rules for FDs and MVDs separately that we have already seen, along with two "mixed" rules, that account for the interaction of FDs and MVDs.

The analogue of closure of an attribute set X is now not an attribute set but rather a collection of attribute sets: $rhs_\Sigma(X) = \{Y\colon\ \Sigma \models X \twoheadrightarrow Y\}$. Now $rhs_\Sigma(X)$ can contain exponentially many sets, and hence is not very useful algorithmically. However, using the system $\mathcal{MVD}$ it is not hard to verify that $rhs_\Sigma(X)$ is a Boolean algebra. Furthermore, since it is a field of finite sets, it is an *atomic* Boolean algebra, and every every element is the union of the atoms it contains. The set of atoms of this Boolean algebra is called the *dependency basis* of X with respect to Σ, denoted $dep_\Sigma(X)$. Thus $dep_\Sigma(X) = \{Y\colon\ Y \neq \emptyset,\ \Sigma \models X \twoheadrightarrow Y$, and if $\Sigma \models X \twoheadrightarrow Z,\ Z \subseteq Y$, and $Z \neq \emptyset$, then $Z = Y\}$.

Lemma 3.2.4. [BFH] $dep_\Sigma(X)$ is a partition of U. Furthermore, $\Sigma \models X \twoheadrightarrow Y$ if and only if there are sets $W_1,\ldots,W_m$ in $dep_\Sigma(X)$ such that $Y = W_1 \cup \ldots \cup W_m$.

Beeri [Bee] has shown how $dep_\Sigma(X)$ can be constructed efficiently using the system $\mathcal{MVD}$.

Theorem 3.2.5. [Bee] The implication problem for MVDs can be solved in time $O(n^4)$, where n is the length of the input. □

Beeri's algorithm was improved by Hagihara et al. [HITK], Sagiv [Sag1], and finally by Galil [Ga]. Galil's algorithm runs in time $O(n \log n)$. These papers and [Bee,BFH] discuss also the interaction of FDs and MVDs.

It is easy to see that Lemma 3.2.4 does not depend on Σ being a set of MVDs. Thus, testing whether an MVD $X \twoheadrightarrow Y$ is implied by a set Σ of dependencies can be done efficiently as long as $dep_\Sigma(X)$ can be constructed efficiently. This was shown in [MSY,Va4] to be the case when Σ is a set of JDs and FDs, and in [Va1] for the case when Σ is a set of typed full dependencies.

Theorem 3.2.6. [Va1] Testing whether an MVD or an FD is implied by a set of typed full dependencies can be done in time $O(n^2)$, where n is the length of the input. □

Let us refer now to implication of JDs. Aho et al. [ABU] described an algorithm, called later the *chase*, to test implication of JDs by FDs.

Theorem 3.2.7. [ABU] Testing whether a JD is implied by a set of FDs can be done in time $O(n^4)$, where n is the length of the input. □

More efficient implementations of the chase were described by Liu and Demers [LD] and by Downey et al. [DST]. The latter algorithm runs in time $O(n^2 \log^2 n)$.

The ideas in [ABU] were generalized by Maier et al. [MMS] to deal with arbitrary implication of FDs and JDs.

Theorem 3.2.8. [MMS] The implication problem for FDs and JDs is solvable in time $O(n^n)$, where n is the length of the input. □

The question then arose whether the exponential upper bound of the above theorem can be improved. Unfortunately, Theorems 3.2.6 and 3.2.7 probably describe the most general case for which an efficient decision procedure exists. Recall that a problem is NP-hard if it is as hard as any problem that can be solved in nondeterministic polynomial time. A problem is NP-complete if it is NP-hard and it can be solved in nondeterministic polynomial time. It is believed that NP-hard problems can not be solved efficiently, i.e., in polynomial time. ([GJ] is a good textbook on the theory of NP-completeness.) Thus, proving that a problem is NP-hard is a strong indication that the problem is computationally intractable.

Theorem 3.2.9.
1) [FT] Testing whether a set of MVDs implies a JD is NP-hard.
2) [BV3] Testing whether a JD and an FD imply a JD is NP-complete. □

Thus, we know how to test implication of FDs and JDs in exponential time, and we know that the problem is NP-hard. We do not know, however, how to pinpoint the complexity of this problem. We do not know for example whether testing implication of a JD by a set of MVDs can be done in nondeterministic polynomial time. One approach to the problem was to try to find a axiom system for FDs and JDs. Surprisingly, even for JDs alone finding a axiom system is extremely difficult (see [BV1,BV5,Sc3]).

NP-completeness strongly suggests, rather than proves, that a problem is intractable (i.e., it proves intractability under the assumption that there are problems that can be solved in nondeterministic polynomial time but not in deterministic polynomial time). In contrast, EXPTIME-completeness is a proof that a problem is intractable. A problem is EXPTIME-complete if it can be solved in exponential time and it is also as hard as any problem that can be solved in exponential time. Since it is known that there are problems that can be solved in exponential time and in fact do require exponential time, it follows that EXPTIME-complete problems require exponential time.

Theorem 3.2.10. [CLM2] The implication problem for typed full dependencies is EXPTIME-complete. □

Interestingly, Beeri and Vardi presented an elegant axiom system for typed full dependencies [BV4]. This demonstrates that there is no clear relationship between having an axiom system for a class of dependencies and the complexity of the implication problem for that class.

In conclusion of this section, the reader should keep in mind that the above lower bounds describe a worst-case behavior of the problems. It is not clear at all that this worst-case behavior indeed arise in practice. We shall return to this point in Section 5.

3.3. The implication problem for embedded dependencies

While for full dependencies the implication problem is clearly solvable and the questions to answer involve upper and lower bounds, this is not so with embedded dependencies, since satisfiability and finite satisfiability do not coincide for the class of $\exists^*\forall^*\exists^*$ sentences, and the corresponding problems are both unsolvable [DG]. Thus, we have to deal here with both implication and finite implication and their corresponding decision problem. Since the class of dependencies is a proper subset of the class of $\forall^*\exists^*$ sentences, one may hope that the (finite) implication problem for embedded dependencies is solvable.

The first disappointing observation is that implication and finite implication do not coincide for embedded dependencies.

Theorem 3.3.1. [CFP,JK] There is a set Σ of FDs and INDs and a single IND τ such that $\Sigma \models_f \tau$, but $\Sigma \not\models \tau$.

Proof: (a) Let Σ be $\{A \rightarrow B, A \subseteq B\}$, and let τ be $B \subseteq A$. We first show that $\Sigma \models_f \tau$. Let I be a finite relation satisfying Σ. We now show that I satisfies τ, that is, $I[B] \subseteq I[A]$. Since I satisfies $A \rightarrow B$ it follows that $|I[B]| \leq |I[A]|$. Since $I[A] \subseteq I[B]$, it follows that $|I[A]| \leq |I[B]|$. Thus, $|I[A]| = |I[B]|$. But since $I[A] \subseteq I[B]$ and since both $I[A]$ and $I[B]$ are finite, we than have $I[B] = I[A]$, so $I[B] \subseteq I[A]$. This was to be shown.

To show that $\Sigma \not\models \tau$, we need only exhibit a relation (necessarily infinite) that satisfies Σ but not τ. Let I be the relation with tuples $\{(i+1,i): i \geq 0\}$. It is obvious that I satisfies Σ but not τ. □

One may think that this behavior is the result of the interaction between tuple-generating dependencies and equality-generating dependencies, but an example in [BV7] shows that even for tuple-generating dependencies the two notions of implication and finite implication differ.

The simplest instance of embedded dependencies are the EMVDs. The (finite) implication problem for EMVDs has resisted efforts of many researchers, and is one of the most outstanding open problems in dependency theory. A significant part of the research in this area has been motivated by this problem. For example, underlying the search for bigger and bigger classes of dependencies was the hope that for the larger class a decision procedure would be apparent, while the specialization of the algorithm to EMVDs was too murky to be visible. Also, underlying the work on axiomatization was the hope that an axiom system may lead to a decision procedure just as the axiom systems for FDs and MVDs led to decision procedures for these classes of dependencies.

Maier et. al [MMS] suggested an extension of the chase to deal with EJDs, and this was further generalized by Beeri and Vardi [BV2] to arbitrary dependencies. Unfortunately, the chase may not terminate for embedded dependencies. It was shown, however, that the chase is a *proof procedure* for implication. That is, given Σ and τ, the chase will give a positive answer if $\Sigma \models \tau$, but will not terminate if $\Sigma \not\models \tau$. Furthermore, Beeri and Vardi [BV4] also presented a sound and complete axiom system for typed dependencies. Nevertheless, all these did not seem to lead to a decision procedure for implication. In 1980 researchers started suspecting that the (finite) implication problem for embedded dependencies was unsolvable, and the first result in this direction were announced in June 1980 by two independent teams.

Theorem 3.3.2. [BV6,CLM1] The implication and the finite implication problem for tuple-generating dependencies are unsolvable. □

This result is disappointing especially with regard to finite implication, which is the more interesting notion. As we recall, $\not\models_f$ is recursively enumerable. Thus, if $\models_f$ is not recursive, then it is not even recursively enumerable. That means that there is no sound and complete axiom system for finite implication.

Both proofs of Theorem 3.3.2 seem to use *untypedness* in a very strong way, and do not carry over to the typed case. Shortly later, however, both teams succeeded in ingeniously encoding untyped dependencies by typed dependencies.

Theorem 3.3.3. [BV7,CLM2] The implication and the finite implication problem for typed tuple-generating dependencies are unsolvable. □

As dependencies, EMVDs have four important properties (see for example (2.3)):

(1) they are tuple-generating,

(2) they are typed,

(3) they have a single atomic formula on the right-hand side of the implication, and

(4) they have two atomic formulas on the left-hand side of the implication.

Dependencies that satisfy properties (1), (2), and (3) above are called *template dependencies,* or TDs [SU]. Thus, EMVDs and EJDs are in particular TDs. Since Theorem 3.3.2 covers properties (1) and (2), the next step was to extend unsolvability to TDs.

Theorem 3.3.4. [GL,Va2] The implication and finite implication problems for TDs are unsolvable. □

In fact, Vardi [Va2] proved a stronger result: the unsolvability for the class of *projected join dependencies.* A projected join dependency (PJD) is of the form $\bowtie[X_1,...,X_k]_X$, where $X \subseteq X_1 \cup ... \cup X_k \subseteq U$. It is obeyed by a relation I if $I[X] = \bowtie\{I[X_1],...,I[X_k]\}[X]$. For an application of PJDs see [MUV]. PJDs extend slightly JDs, since if $X = X_1 \cup ... \cup X_k$, then the PJD $\bowtie[X_1,...,X_k]_X$ is equivalent to the JD $\bowtie[X_1,...,X_k]$. Thus the class of PJDs lies strictly between the classes of EJDs and TDs. The implication and finite implication problems for EJDs are, however, still wide open.

Even though the existence of an axiom system for a certain class of dependencies does not guarantee solvability of the implication problem, finding such a system seems to be a valuable goal. In particular attention was given to *k-ary* systems. In a k-ary axiom systems, all inference rules are of the form $\tau_1,...,\tau_n \vdash \tau$, where $n \leq k$. It is easy to verify, for example, that the systems $\mathcal{FD}$ and $\mathcal{MVD}$ in Section 3.3 are 2-ary.

Theorem 3.3.5. [PP,SW] For all $k>0$, there is no sound and complete k-ary axiom system for implication and finite implication of EMVDs. □

We refer the reader to [BV4,Va2] for a discussion regarding the existence of a non-k-ary axiom system for EMVDs.

Let us refer now to what some people believe are the only "practical" dependencies, FDs and INDs.

Recall that FDs are full dependencies, so implication and finite implication coincide and both are solvable (and by Theorem 3.2.1, quite efficiently). INDs, on the other hand, are embedded dependencies, so a straightforward application of the chase does not yield a decision procedure. A more careful analysis, however, shows that the chase can be forced to terminate.

Theorem 3.3.6. [CFP] The implication and finite implication problem for INDs are equivalent and are PSPACE-complete. □

(PSPACE-complete problems are problems that can be solved using only polynomial space and are hard as any problem that can be solved using polynomial space. It is believed that this problems can not be solved in polynomial time [GJ].)

Let us consider now implication of arbitrary dependencies by IND· Since containment of tableaux [ASU] can be expressed by dependencies [YP], a test for implication of dependencies by INDs is also a test for containment of conjunctive queries under INDs. We do not know whether implication and finite implication coincide in this case. We have, however, a positive result for implication.

Theorem 3.3.7. [JK] Testing implication of dependencies by INDs is PSPACE-complete. □

The finite implication problem for this case is still open.

Casanova et al. [CFP] investigated the interaction of FDs and INDs, and they discovered that things get more complicated when both kinds of dependencies are put together. First, they showed that implication and finite implication are different (Theorem 3.3.1). In addition they showed that there is no sound and complete k-ary axiom system for implication and finite implication of FDs and INDs. (Mitchell [Mi1], however, has shown that in a more general sense there is a k-ary axiom system for implication of FDs and INDs.) In view of their results, it did not come as a surprise when Chandra and Vardi and, independently, Mitchell proved unsolvability.

Theorem 3.3.8. [CV,Mi2] The implication and the finite implication problems for FDs and INDs are unsolvable. □

Some people claim is that in practice we encounter only INDs that have a single attribute on each side of the containment, e.g., MANAGER⊆EMPLOYEE. Such INDs are called *unary* INDs (UINDs). Reviewing the proof of Theorem 3.3.1, we realize that even for FDs and UINDs implication and finite implication differ. Considering our experience with dependencies, this looks like a sure sign that the problems are unsolvable. The next result by Kannelakis et al. comes therefore as a refreshing surprise.

Theorem 3.3.9. [KCV] The implication and the finite implication problem for FDs and UINDs are both solvable in polynomial time. □

For other positive results for INDs see [KCV,JK,LMG].

In conclusion to this topic, we would like to mention an argument against the relevance of all the above unsolvability results. The assumption underlying these results is that the input is an arbitrary set Σ of dependencies and a dependency τ. The argument is that the given set Σ is supposed to describe some "real life" application, and in practice it is not going to be arbitrary. Thus, even if we concede that TDs arise in practice, still not every set of TDs arises in practice. The emphasis of this argument is on "real world sets of dependencies", rather than on "real world dependencies". For further study of this argument see [Sc1,Sc2]. While we agree with the essence of this argument, we believe that the results described above are useful in delineating the boundaries between the computationally feasible and infeasible. This is especially important, since we do not yet have robust definitions of real world sets of dependencies.

4. The Universal Relation Model

4.1. Motivation

A primary justification given by Codd for the introduction of the relational model was his view that earlier models were not adequate to the task of boosting the productivity of programmers [Co1,Co3]. One of his stated motivations was to free the application programmer and the end user from the need to specify access paths (the so-called "navigation problem"). A second motivation was to eliminate the need for program modification to accommodate changes in the database structure, i.e., to eliminate access path dependence in programs.

After a few years of experience with relational database management systems, it was realized [CK] that, though being a significant step forward, the relational model by itself fails to achieve complete freedom from user-supplied navigation and from access path dependence. The relational model was successful in removing the need for *physical navigation*; no access paths need to be specified within the storage structure of a single relation. Nevertheless, the relational model has not yet provided independence from *logical navigation,* since access paths among several relations must still be satisfied.

For example, consider a database that has relations *ED*(Employee, Department) and *DM*(Department, Manager). If we are interested in the relationship between employees and managers through departments, then we have to tell the system to take the join of the *ED* and *DM* relations and to project it on the attributes *EM*. This is of course an access path specification, and if the database were to be reorganized to have a single relation *EDM*, then any programs using this access path would have to be modified accordingly.

The *universal relation model* aims at achieving complete access path independence by letting us ask the system in an appropriate language "tell us about employees and their managers", expecting the system to figure out the intended access path for itself. Of course, we cannot expect the system to always select the intended relationship between employees and managers automatically, because the user might have something other than the simplest relationship, the one through departments, in mind, e.g., the manager of the manager of the employee. We shall, in a universal relation system, have to settle for eliminating the need for logical navigation along certain paths, those selected by the designer, while allowing the user to navigate explicitly in more convoluted ways.

Unlike the relational model, the universal relation model was not introduced as a single clearly defined model, but rather evolved during the 1970's through the work of several researchers. As a result, there have been a significant confusion with regard to the assumptions underlying the model, the so-called "universal relation assumptions". We refer the reader to [MUV], where an attempt is made to clarify the situation.

In this and the next section we restrict ourselves to finite databases.

4.2. Decomposition

The simplest way to implement the universal relation model is to have the database consist a *universal relation,* i.e., a single relation over the set U of all attributes. There are two problems with this approach. First, it assumes that for each tuple in the database we always can supply values for all the attributes, e.g., it assumes that we have full biographic information on all employees. Secondly, storing all the information in one universal relation causes problems when this information needs to be updated. These problems, called *update anomalies,* were identified by Codd [Co2]. The solution to these problems is to have a conceptual database that consists of the universal relation, while the actual database consists of relations over smaller sets of attributes. That is, the database scheme consists of a collection $\mathbf{R} = \{R_1,...,R_k\}$ of attributes sets whose union is U, and the database consists of relations $I_1,...,I_k$, over $R_1,...,R_k$, respectively.

A principal activity in relational database design is the decomposition of the universal relation scheme into a database scheme that has certain nice properties, traditionally called *normal forms.* (We shall not go here into *normalization theory,* which is the study of these normal forms, and the interested reader is referred to [Ma,Ul].) More precisely, starting with the universal scheme U and a set of dependencies Σ, we wish to replace the universal scheme by a database scheme $\mathbf{R} = \{R_1,...,R_k\}$. The idea is to replace the universal relation by its projection on $R_1,...,R_k$. That is, instead of storing a relation I over U, we *decompose* it into $I_1 = I[R_1]$, ..., $I_k = I[R_k]$, and store the result of this decomposition. The map $\Delta_{\mathbf{R}}$ defined by $\Delta_{\mathbf{R}}(I) = \{I[R_1],...,I[R_k]\}$ is called the *decomposition map.*

Clearly, a decomposition cannot be useful unless no loss of information is incurred by decomposing the universal relation. (This is called in [BBG] the *representation principle.*) That is, we must be able to reconstruct I from $I_1,...,I_k$. More precisely, the decomposition map has to be *injective.* For our purposes it suffices that the decomposition map is injective for relations that satisfy the given set Σ of dependencies. In this case we say that it is injective with respect to Σ. When the decomposition map is injective it has a left inverse, called the *reconstruction map.* The basic problems of *decomposition theory* are to formulate necessary and sufficient conditions for injectiveness and to find out about the reconstruction map.

The natural candidate for the reconstruction map is the join, i.e., $I=I_1 \bowtie \ldots \bowtie I_k$. The naturalness of the join led many researchers to the belief that if the reconstruction map exists then it is necessarily the join. This belief was refuted by Vardi [Va3], who constructed an example where the decomposition map is injective, but the reconstruction map is not the join. It is also shown in [Va3] how to express injectiveness as a statement about implication of dependencies. Unfortunately, even when Σ consists of full dependencies, that statement involves also inclusion dependencies. It is not known whether there is an effective test for injectiveness.

If we insist that the join be the reconstruction map, then we can get a stronger result.

Theorem 4.2.1. [BR,MMSU] Let Σ be a set of dependencies, and let **R** be a database scheme. $\Delta_{\mathbf{R}}$ is injective with respect to Σ with the join as the reconstruction map if and only if $\Sigma \models \bowtie[\mathbf{R}]$. □

Thus, if Σ consists of full dependencies then we can effectively test whether the decomposition map is injective with respect to Σ.

Another desirable property of decompositions is *independence* [Ri1]. Intuitively, independence means that the relations of the database can be updated independently from each other. For further investigation of the relationship between injectiveness and independence see [BH, BR, MMSU, Va3 .

A point that should be brought up is that decomposition may have some disadvantages. Essentially, decomposition may make it easier to update the database, but it clearly makes it harder to query it. Since the join operation can be quite expensive computationally, reconstructing the universal relation may not be easy even when the reconstruction map is the join. In fact, even testing whether the relations of the database can be joined without losing tuples is NP-complete, and hence, probably computationally intractable. Let the database consists of relations $I_1,...,I_k$ over attribute sets $R_1,...,R_k$. We say that the database is *join consistent* if there is a universal relation I such that $I_j = I[R_j]$, for $1 \leq j \leq k$. (Rissanen [Ri1] calls a join consistent set of relations *joinable*. A join consistent database is also called *globally consistent* [BFMY], *join compatible* [BR], *valid* [Ri3], *consistent* [Fa6], or *decomposed* [Va3].) It is easy to verify that the database is join consistent if $I_j = \bowtie \{I_1,...,I_k\}[R_j]$, for $1 \leq j \leq k$.

Theorem 4.2.2. [HLY] Testing whether a database is join consistent is NP-complete. □

Thus there is a trade-off between the ease of updating the database and the ease of querying it. The smaller the relation schemes, the easier it is to update the database and the harder it is to query it. Recognizing this trade-off, Schkolnick and Sorenson investigated what they called *denormalization* [SS]. The idea is to decompose the universal scheme with both the ease of updating and the ease of querying in mind. The result of the decomposition depends in this approach on the predicted use of the database.

4.3. The Universal Relation Interface

Suppose now that decomposition has been achieved. That is, assume that, starting with the universal scheme U and a set Σ of dependencies, we have designed a database scheme $\mathbf{R} = \{R_1,...,R_k\}$, and we now have a database $\mathbf{I} = \{I_1,..,I_k\}$ over $\mathbf{R}$. Two questions have now to be resolved: how to determine whether the database is semantically meaningful, i.e., satisfies the given dependencies, and how to respond to the users' queries that refer to the universal relation. If the database is join consistent, then we can construct the universal relation I such that $\Delta_{\mathbf{R}}(I) = \mathbf{I}$. But if the database is not join consistent, then there is no corresponding universal relation.

We outline here one approach to the problem, called the *weak* universal relation approach. (This approach was suggested by Honeyman [Ho] and further developed in [GM,GMV,MUV]. For other approaches and their relationship to the weak universal relation approach see [GM,GMV,MRW,MUV].) According to this approach, a universal relation exists at least in principle, even though it may not be known. The database is seen, from this viewpoint, as a partial specification of the universal relation. More precisely, the relations $I_1,...,I_k$ are partial descriptions of the projection of the universal relation I on the relation schemes $R_1,...,R_k$. Thus a universal relation I is considered to be a *weak universal relation* for **I** with respect to Σ if it satisfies Σ and $I_j \subseteq I[R_j]$, for $1 \leq i \leq k$. **I** is *consistent* with Σ i.e., semantically meaningful, if it has a weak universal relation with respect to Σ.

The above definition is existential in nature and does not lend itself to an effective test. The *consistency problem* is to decide, for a given set Σ of dependencies and a database **I** over a database scheme **R**, whether **I** is consistent with Σ.

Theorem 4.3.1.
1) [GMV] The consistency problem for embedded dependencies is unsolvable.
2) [GMV] The consistency problem for full dependencies is EXPTIME-complete.
3) [Ho] The consistency problem for FDs is solvable in polynomial time. □

Thus, for embedded dependencies there is no effective test for consistency, for full dependencies there is an effective though intractable test, and the good news is that for FDs there is a polynomial time test for consistency. We note that the presence of the independence property, mentioned is Section 4.2, may make it easier to test for consistency. We refer the reader to [CM,Gr2,GY,Sa] for the study of independence in the context of the weak universal relation approach.

We now refer to the issue of query answering. For simplicity we restrict ourselves to queries of the form "give me the relationship between employees and managers". More precisely, the query is a set X of attributes, and the desired answer is the so-called *basic relationship on X*. If we had a unique universal relation I, then answer would undoubtedly be $I[X]$. But in our case we have only weak universal relations, and we clearly have infinitely many of those. Since we cannot know which of the possible universal relations actually represent the "real world" at a given moment, we assume that the only facts that can be deduced about the universal relation from the database are those that hold is all weak universal relations. This motivated researchers ([MUV,Ya] following [Sa]) to adopt the following definition. Let *weak*(I,Σ) be the set of all weak universal relations of I with respect to Σ. We can see this set as the embodiment of the information represented by the database [Me]. The answer to the query X, denoted I[X], is therefore taken to be $\cap \{I[X] : I \epsilon weak(\mathbf{I},\Sigma)\}$. Note that the answer is with respect to Σ.

The above definition does not seem to lead to an effective procedure for computing I[X].

Theorem 4.3.2.
1) [GMV] Computing answers with respect to embedded dependencies is unsolvable.
2) [GMV] Computing answers with respect to full dependencies is EXPTIME-complete.
3) [Ho] Computing answers with respect to FDs can be done in polynomial time. □

We refer the reader to [Gr3,MRW,MUV,Sag2,Sag3,Ya] for further study of query answering.

We conclude this section by considering again the questions raised in the previous section. There we started with a universal relation I and applied the decomposition map $\Delta_{\mathbf{R}}$, to get the database $\Delta_{\mathbf{R}}(I) = \{I[R_1],...,I[R_k]\}$. Suppose now that we pose the query U to this database. In this case we would expect our query answering mechanism to be the desired reconstruction map, i.e., we would expect $I = \Delta_{\mathbf{R}}(I)[U]$.

Theorem 4.3.3. [MUV] The following two conditions are equivalent:
1) $\Sigma \models \bowtie [\mathbf{R}]$.
2) $I = \Delta_{\mathbf{R}}(I)[U]$, for every universal relation I that satisfies Σ. □

In other words, if our query answering mechanism happens to be the reconstruction map, then for join consistent databases it is actually the join.

5. Acyclic database schemes

In the last few years, acyclic database schemes have been introduced and studied [BFMMUY]. The idea is to view database schemes as hypergraphs. A *hypergraph* is a pair $(\mathscr{N}, \mathscr{E})$, where $\mathscr{N}$ is a finite set of *nodes*, and $\mathscr{E}$ is a set of *edges* (or *hyperedges*), which are arbitrary nonempty subsets of $\mathscr{N}$. An ordinary undirected graph (without self-loops) is, of course, a hypergraph where every edge has exactly two nodes. We shall identify a hypergraph by only mentioning its edges, and tacitly assume that the nodes are precisely those that appear in some edge. We can then view a database scheme as a hypergraph where the relation schemes are the edges. The hypergraph of Figure 5.2 corresponds to the database scheme of Figure 5.1. The correspondence should be clear: for example, there is a {SUPPLIER,PART,COST} edge in the hypergraph of Figure 5.2 because of the {SUPPLIER,PART,COST} relation scheme in Figure 5.1, and so on.

Unlike the situation for ordinary undirected graphs, there are a number of inequivalent, natural definitions of acyclicity for hypergraphs. The type of acyclicity (due to [BFMMUY]) which we shall focus on in this section can be defined by a generalization of one of the usual definitions of acyclicity for ordinary undirected graphs. In particular, an ordinary undirected graph is acyclic in the usual sense if and only if, when considered as a hypergraph, it is acyclic under our definition.

A number of basic, desirable properties of relational database schemes turn out to be equivalent to acyclicity. These properties were defined and studied by a number of researchers, in quite different contexts. It is somewhat remarkable that all of these properties are equivalent. We describe here only a few of these properties. For a more complete survey, see [BFMY, Fa5, Fa6]. Furthermore, there are arguments that the class of acyclic database schemes are natural from a semantic point of view [Li, Sc1].

There is a simple, efficient algorithm for determining acyclicity. For this paper, we shall simply take this algorithm as defining acyclicity. The algorithm is called *Graham's algorithm*, in honor of Marc Graham, who showed [Gr1] that if a hypergraph was accepted by his algorithm, then this was sufficient to imply a certain nice database property. Graham's algorithm was also defined, independently, by Yu and Ozsoyoglu [YO].

The algorithm proceeds by applying the following operations repeatedly, in any order, until none can be applied:

(a) if a node is isolated (that is, if it belongs to precisely one edge), then delete that node;

(b) if an edge is a subset of another edge, then delete the first edge.

The algorithm clearly terminates. If the end result is the empty set, then the original hypergraph is acyclic; otherwise, it is cyclic.

As an example, let us apply this algorithm to the hypergraph of Figure 5.3. Somewhat surprisingly, it turns out that this hypergraph is acyclic, even though it seems to contain a "cycle"; we shall come back to this point later.

We begin the algorithm by writing the edges, one underneath the other:

A	B	C			
		C	D	E	
A				E	F
A		C		E	

(For convenience, we have put common vertices in the same column.)

We begin by deleting the isolated nodes B, D, and F. We are left with:

A	C	
	C	E
A		E
A	C	E

Since the first (AC) row is contained in the last (ACE) row, we now delete the first row:

	C	E
A		E
A	C	E

We now delete the new first and second rows, since each is contained in the new third row. We are left with a single row:

A		C		E

We now delete the isolated nodes A, C, and E. We are then left with the empty set. Since the algorithm terminates with the empty set, the hypergraph of Figure 5.3 is acyclic.

It is instructive to see an example in which the hypergraph is cyclic. This time, we apply the algorithm to the hypergraph of Figure 5.4. This hypergraph contains three of the four edges of the hypergraph of Figure 5.3. The algorithm begins with

A	B	C			
		C	D	E	
A				E	F

After deleting the isolated nodes B, D, and F, we are left with:

A		C		
		C		E
A				E

The algorithm now halts, since no node is isolated and no row is a subset of another row. Since what is left is not the empty set, the hypergraph is cyclic.

Note that the acyclic hypergraph of Figure 5.3 has a cyclic subhypergraph, namely, the hypergraph of Figure 5.4. (A *subhypergraph* of a hypergraph is simply the hypergraph consisting of a subset of the edges.) This counterintuitive phenomenon does not happen with ordinary graphs: that is, it is not possible for a subgraph of an ordinary acyclic graph to be cyclic. Later, we shall mention another type of acyclicity for hypergraphs, where this counterintuitive phenomenon does not occur.

A simple analysis of a natural implementation of Graham's algorithm shows that it can made to run in cubic time. We remark that Tarjan and Yannakakis [TY] have recently obtained a a linear time algorithm for determining acyclicity.

We now discuss a particular desirable property of database schemes that is equivalent to acyclicity. Recall that a database is *join consistent* if there is a single universal relation such that each relation in the database is the appropriate projection of U.

We say that a pair of relations is join consistent if the database consisting only of these two relations is join consistent. Let us say that the two relations are I_1, with attributes R_1, and I_2, with attributes R_2. Let $X = R_1 \cap R_2$. Thus, X is the set of attributes that I_1 and I_2 have in common. It is easy to see that I_1 and I_2 are join consistent precisely if $I_1[X] = I_2[X]$, that is if they agree on their common part. Let us say that a database is *pairwise consistent* if each pair of relations is join consistent. It is clear every join consistent database is pairwise consistent. It would be very nice if the converse were true, since then there would be a simple test for join consistency, namely, pairwise consistency. Unfortunately, however, the converse does not hold. For, it is easy to verify that the database of Figure 5.5 is pairwise consistent; however, it is not hard to see that it is not join consistent. In fact, we already knew that there could be no simple test for join consistency, since, as we noted earlier, determining join consistency is an NP-complete problem [HLY].

However, in the acyclic case our desired converse holds, that is, pairwise consistency and join consistency are equivalent.

Theorem 5.1. [BFMY] If the scneme is acyclic, then a database is join consistent if and only if it is pairwise consistent. □

Can there be any *cyclic* schemes for which join consistency and pairwise consistency are equivalent? The answer is no.

Theorem 5.2. [BFMY] If the scheme is cyclic, then there is a database that is pairwise consistent but not join consistent. □

Putting Theorems 5.1 and 5.2 together, we see that a scheme is acyclic if and only if every pairwise consistent database is join consistent. Therefore (using also the fact that join consistency implies pairwise consistency), it follows that a scheme is acyclic if and only if checking pairwise consistency is an algorithm for testing join consistency. This is an example of a desirable database property that is equivalent to acyclicity.

There is one viewpoint on what we have just discussed that should be emphasized. In the *unrestricted* case (where we do not assume acyclicity), testing join consistency is an NP-complete problem. However, in the *acyclic* case, there is a polynomial-time algorithm for testing join consistency (namely, testing pairwise consistency). This gives us an example of an NP-hard problem that has a polynomial-time algorithm if the scheme is acyclic. We shall mention another such example soon.

We now consider another condition which is equivalent to acyclicity of a hypergraph. The join dependency $\bowtie [X_1, ..., X_k]$ is said to be acyclic precisely if the hypergraph with edges $X_1, ..., X_n$ is acyclic.

Theorem 5.3. [FMU] A join dependency is acyclic if and only if it is equivalent to a set of multivalued dependencies. □

It is known [BFMY] that an acyclic join dependency $\bowtie [R_1, ..., R_n]$ is in fact equivalent to a set of at most n-1 multivalued dependencies (where n is the number of R_i's). Further, the constructions in [BFMY] show that there is a polynomial-time algorithm for finding such a set of n-1 multivalued dependencies. As an example, the acyclic join dependency $\bowtie [ABC, CDE, EFA, ACE]$, which corresponds to the acyclic hypergraph of Figure 5.3, is equivalent to the set $\{AC \twoheadrightarrow DEF, CE \twoheadrightarrow ABF, AE \twoheadrightarrow BCD\}$ of multivalued dependencies.

Beeri et al. [BFMY] and Goodman and Tay [GT] consider the converse question, of when a given set of MVDs is equivalent to some JD. In particular, Goodman and Tay give a polynomial-time algorithm for answering this question.

We can now give another example of an NP-hard problem which acyclicity rends tractable. The problem of deciding whether a set of typed full dependencies implies a JD is NP-hard; in fact, as noted in Theorem 3.2.9 above, this is even true if all of the typed full dependencies are MVDs. However, if the join dependency is acyclic, then there is a polynomial-time algorithm.

Theorem 5.4. Testing whether a set of typed full dependencies implies an acyclic JD can be done in polynomial time.

Proof: As noted above, there is a polynomial-time procedure for finding a set of n-1 MVDs which are equivalent to the given JD $\bowtie [R_1, ..., R_n]$. By Theorem 3.2.8, there is a polynomial-time algorithm for deciding whether the set Σ of typed full dependencies implies each of the multivalued dependencies. Clearly, Σ implies the acyclic JD if and only if it implies all of these MVDs. □

We note that Yannakakis [Ya] has found other problems involving JDs that are NP-hard in general, but which have polynomial-time algorithms if the JDs are acyclic.

Our final example for the desirability of acyclicity involves the property of injectiveness discussed in Section 4.2. We noted there that injectiveness of a decomposition map does not guarantee that the reconstruction map is the join. The situation is different if the decomposition is acyclic.

Theorem 5.5. [BV8] Let Σ be a set of full dependencies, and let **R** be an acyclic database scheme. If $\Delta_{\mathbf{R}}$ is injective with respect to Σ, then the reconstruction map is the join. □

Note that combining Theorems 4.2.1, 5.4, and 5.5, we get a polynomial time test for the injectiveness of an acyclic decomposition with respect to typed full dependencies.

The idea of acyclicity has turned out to be a powerful unifying concept. We refer the reader to [BC, BG, GS] for applications to query processing, which we do not discuss here. We note that Sacca [Sa] and Laver et al. [LMG] deal with the effect of FDs on acyclicity. Further, Biskup and Brüggemann [BiBr] study the effect of FDs on the design process of acyclic database schemes.

Fagin [Fa6] has introduced even more restrictive types of acyclicity, which correspond to database schemes that enjoy even nicer properties. We now focus on one such type of acyclicity, called *γ-acyclicity*. For convenience, let us refer (as Fagin does in [Fa6]) to the type of acyclicity we have been discussing as *α-acyclicity*. As we noted earlier, the definition of α-acyclicity has the counterintuitive property that a subhypergraph of an α-acyclic hypergraph may be α-cyclic. As an example, the α-acyclic hypergraph of Figure 5.3 has the α-cyclic subhypergraph of Figure 5.4. It turns out that for γ-acyclicity, this counterintuitive phenomenon does not occur; thus, every subhypergraph of a γ-acyclic hypergraph is γ-acyclic.

There are various graph-theoretic definitions of γ-acyclicity [Fa6]. For the purposes of this paper, we simply give an algorithm for determining γ-acyclicity. This algorithm is due to D'Atri and Moscarini [DM], and was proven correct by Fagin [Fa6]. It is very similar in spirit to Graham's algorithm for α-acyclicity, which we presented earlier. We then apply the following operations repeatedly, in any order, until none can be applied:

(a) if a node is isolated (that is, if it belongs to precisely one edge), then delete that node;

(b) if an edge is a singleton (that is, if it contains exactly one node), then delete that edge (but do not delete the node from other edges that might contain it);

(c) if an edge is empty, then delete it;

(d) if two edges contain precisely the same nodes, then delete one of these edges;

(e) if two nodes are edge-equivalent, then delete one of them from every edge that contains it. (We say that two nodes are *edge-equivalent* if they are in precisely the same edges.)

The algorithm clearly terminates. If the end result is the empty set of edges, then the original hypergraph is γ-acyclic; otherwise, it is γ-cyclic.

As an example, let us apply this algorithm to the hypergraph of Figure 5.6. As in the previous example, we begin by by writing the edges, one underneath the other. The edges are:

	B	C	D	E	F
A	B	C	D		
		C			
		C	D		
				E	F

Node A is isolated, and edge $\{C\}$ is a singleton, so both are deleted, by rules (a) and (b). This leaves us with

B	C	D	E	F
B	C	D		
	C	D		
			E	F

Nodes E and F are edge-equivalent, and so, by rule (e), we delete F from both edges that contain it. Similarly, nodes C and D are edge-equivalent, and so we delete D from all three edges that contain it. We are left with

B	C	E
B	C	
	C	
		E

The third and fourth edges above are singletons, and so they are eliminated. This leaves

B	C	E
B	C	

Node E is isolated; after it is deleted, we are left with

B	C
B	C

These edges are identical, so we delete one by rule (d). We are left with

$$B \quad C$$

Both nodes are now isolated, and so they are deleted. We are left with a single empty edge, which is deleted by rule (c). The end result is the empty set of edges, and so the original hypergraph is γ-acyclic.

Fagin [Fa6] presents a number of desirable database properties, each which is equivalent to the scheme being γ-acyclic. We mention only one property, which involves dependencies. We first need to define the concept of connectedness for hypergraphs; it is the obvious generalization of the definition of connectedness for graphs. A *path* from node s to node t is a sequence of $k \geq 1$ edges $E_1, \ldots, E_k$ such that

(i) s is in E_1,
(ii) t is in E_k, and
(iii) $E_i \cap E_{i+1}$ is nonempty if $1 \le i < k$.

A hypergraph is *connected* if for each pair of nodes, there is a path from one to the other.

Theorem 5.5. [Fa6] Let **R** be a database scheme. Then the following are equivalent:

(1) **R** is γ-acyclic.

(2) For every connected subset $\mathbf{S} \subseteq \mathbf{R}$, we have $\bowtie[\mathbf{R}] \models \bowtie[\mathbf{S}]$. □

Note that $\bowtie[\mathbf{S}]$ can be an embedded JD. Thus statement (2) of Theorem 5.5 above says that for every connected subset **S** of **R**, and for every join consistent database **I** over **R**, if $\mathbf{J} \subseteq \mathbf{I}$ is the subdatabase over **S**, then $\bowtie \mathbf{J}$ is a projection of $\bowtie \mathbf{I}$.

As an example, consider the database scheme of Figure 5.1, which is γ-cyclic. The join of the {SUPPLIER,PROJECT,DATE} relation with the {PROJECT,PART,COUNT} relation might introduce a SUPPLIER, PART, PROJECT triple that does not appear in the SUPPLIER, PART, PROJECT relation (the "connection trap" [Co1].)

We note also that the various notions of acyclicity turn out to be very useful for the design of universal relation interfaces. We refer the reader to [Fa6, MU, Ya].

BIBLIOGRAPHY

[ABU] A. V. Aho, C. Beeri, and J. D. Ullman, The theory of joins in relational data bases. ACM Trans. on Database Systems 4(1979), 297-314.

[ASU] A. V. Aho, Y. Sagiv, and J. D. Ullman, Equivalences among relational expressions. SIAM J. Computing 8(1979), 218-246.

[Ar] W. W. Armstrong, Dependency structures of database relationships. Proc. IFIP 74, North Holland, 1974, 580-583.

[Bee] C. Beeri, On the membership problem for functional and multivalued dependencies in relational databases. ACM Trans. on Database Systems 5(1980), 241-259.

[BB] C. Beeri and P. A. Bernstein, Computational problems related to the design of normal form relational schemas. ACM Trans. on Database Systems 4(1979), 30-59.

[BBG] C. Beeri, P. A. Bernstein, and N. Goodman, A sophisticate's introduction to database normalization theory. Proc. Int. Conf. on Very Large Data Bases, 1978, Berlin, 113-124.

[BFH] C. Beeri, R. Fagin, and J.H. Howard, A complete axiomatization for functional and multivalued dependencies in database relations. Proc. ACM SIGMOD Conf. on Management of Data, 1977, Toronto, 47-61.

[BFMMUY] C. Beeri, R. Fagin, D. Maier, A. O. Mendelzon, J. D. Ullman, and M. Yannakakis, Properties of acyclic database schemes. Proc. 13th ACM SIGACT Symp. on Theory of Computing, 1981, Milwaukee, 355-362.

[BFMY] C. Beeri, R. Fagin, D. Maier, and M. Yannakakis, On the desirability of acyclic database schemes. J. ACM 30(1983), 479-513.

[BH] C. Beeri and P. Honeyman, Preserving functional dependencies. SIAM J. Computing 10(1981), 647-656.

[BK] C. Beeri and H. F. Korth, Compatible attributes in a universal relation. Proc. 1st ACM SIGACT-SIGMOD Symp. on Principles of Database Systems, 1982, Los Angeles, 51-62.

[BMSU] C. Beeri, A. O. Mendelzon, Y. Sagiv, and J. D. Ullman, Equivalence of relational database schemes, SIAM J. Computing 10(1981), 352-370.

[BR] C. Beeri and J. Rissanen, Faithful representation of relational database schemes. IBM Research Report, San Jose, California, 1980.

[BV1] C. Beeri and M. Y. Vardi, On the properties of join dependencies. In Advances in Database Theory (H. Gallaire, J. Minker, and J. M. Nicolas, Eds.), Plenum Press, 1981, 25-72.

[BV2] C. Beeri and M. Y. Vardi, A proof procedure for data dependencies. J. ACM 31(1984), 718-741.

[BV3] C. Beeri and M. Y. Vardi, On the complexity of testing implications of data dependencies. Hebrew University of Jerusalem Technical Report, Dec. 1980.

[BV4] C. Beeri and M. Y. Vardi, Formal systems for tuple and equality-generating dependencies. SIAM J. Computing 13(1984), 76-98.

[BV5] C. Beeri and M. Y. Vardi, Formal systems for join dependencies. Hebrew Univ. of Jerusalem Technical Report, 1981. To appear in Theoretical Computer Science.

[BV6] C. Beeri and M. Y. Vardi, The implication problem for data dependencies. Proc. XP1 Workshop on Relational Database Theoery, Stony Brook, NY, June 1980.

[BV7] C. Beeri and M. Y. Vardi, The implication problem for data dependencies. Proc. 8th Int. Colloq. on Languages Automata and Programming, 1981, Acre Israel. Appeared in: Lecture Notes in Computer Science - Vol. 115, Springer-Verlag, 1981, 73-85.

[BV8] C. Beeri and M. Y. Vardi, On acyclic database decompositions. Information and Control 6(1984), 75-84.

[Ber] P. A. Bernstein, Synthesizing third normal form relations from functional dependencies. ACM Trans. on Database Systems 1(1976), 277-298.

[BC] P. A. Bernstein and D. W. Chiu, Using semi-joins to solve relational queries, J. ACM 28(1981), 25-40.

[BG] P. A. Bernstein and N. Goodman, The power of natural semijoins, SIAM J. Computing 10(1981), 751-771.

[BiBr] J. Biskup and H. H. Brüggemann, Towards designing acyclic database schemas, Proc. ONERA-CERT Workshop on Logical Bases for Data Bases, Toulouse, 1982.

[CFP] M. A. Casanova, R. Fagin, and C. Papadimitriou, Inclusion dependencies and their interaction with functional dependencies. J. Computer and System Sciences 28(1984), 29-59.

[CM] E. P. F. Chan and A. O. Mendelzon, Independent and separable database schemes. Proc. 2nd ACM SIGACT-SIGMOD Symp. on Principles of Database Systems, 1983, Atlanta, 288-296.

[CLM1] A. K. Chandra, H. R. Lewis, and J. A. Makowsky, Embedded implicational dependencies and their inference problem. Proc. XP1 Workshop on Relational Database Theoery, Stony Brook, NY, June 1980.

[CLM2] A. K. Chandra, H. R. Lewis, and J. A. Makowsky, Embedded implicational dependencies and their inference problem. Proc. 13th ACM Symp. on Theory of Computing, 1981, Milwaukee, 342-354.

[CV] A. K. Chandra and M. Y. Vardi, The implication problem for functional and inclusion dependencies is undecidable. IBM Research Report RC 9980, May 1983. To appear in SIAM J. Computing.

[Co1] E. F. Codd, A relational model of data for large shared data banks. Comm. ACM 13(1970), 377-387.

[Co2] E. F. Codd, Further normalizatioı of the data base relational model. Courant Computer Science Symposia 6: Data Base Systems, 1972, Prentice Hall, 33-64.

[Co3] E. F. Codd, Relational databases: a practical foundation for productivity. Comm. ACM 25(1982), 109-117.

[DM] A. D'Atri and M. Moscarini, Acyclic hypergraphs: their recognition and top-down vs bottom-up generation. Consiglio Nazionale Delle Ricerche, Istituto di Analisi dei Sistemi ed Informatica, R.29, 1982.

[DP] P. De Bra and J. Paredaens, Conditional dependencies for horizontal decompositions. Proc. 10th Int. Colloq. on Languages Automata and Programming, 1981, Barcelona. Appeared in: Lecture Notes in Computer Science - Vol. 154, Springer-Verlag, 1983, 67-82.

[DST] P. J. Downey, R. Sethi, and R. E. Tarjan, Variations on the common subexpression problem. J. ACM 27(1980), 758-771.

[DG] B. S. Dreben and W. D. Goldfarb, The Decision Problem: Solvable Classes of Quantificational Formulas. Addison Wesley, 1979.

[Fa1] R. Fagin, Multivalued dependencies and a new normal form for relational databases. ACM Trans. on Database Systems 2(1977), 262-278.

[Fa2] R. Fagin, A normal form for relational databases that is based on domains and keys. ACM Trans. on Database Systems 6(1981), 387-415.

[Fa3] R. Fagin, Armstrong databases. Proc. 7th IBM Symp. on Mathematical Foundations of Computer Science, Kanagawa, Japan, May 1982. Also appeared as IBM Research Report RJ3440 (April 1982), San Jose, California.

[Fa4] R. Fagin, Horn clauses and database dependencies. J. ACM 29(1982), 952-985.

[Fa5] R. Fagin, Acyclic database schemes of various degrees: a painless introduction. Proc. CAAP83 8th Colloquium on Trees in Algebra and Programming., Appeared in: Springer-Verlag Lecture Notes in Computer Science - vol. 159, 1983, ,d. G. Ausiello and M. Protasi, 65-89.

[Fa6] R. Fagin, Degrees of acyclicity for hypergraphs and relational database schemes. J. ACM 30(1983), 514-550.

[FMU] R. Fagin, A. O. Mendelzon, and J. D. Ullman, A simplified universal relation assumption and its properties. ACM Trans. on Database Systems 7(1982), 343-360.

[TF] P. C. Fischer and D.-M. Tsou, Whether a set of multivalued dependencies implies a join dependency is NP-hard. SIAM J. Computing 12(1983), 259-266.

[Ga] Z. Galil, An almost linear-time algorithm for computing a dependency basis in a relational database. J. ACM 29(1982), 96-102.

[GJ] M. R. Garey and D. S. Johnson, Computers and Intractibility: A Guide to the Theory of NP-Completeness. Freeman, 1979.

[GZ] S. Ginsburg and S. M. Zaiddan, Properties of functional dependency families. J. ACM 29(1982), 678-698.

[GS] N. Goodman and O. Shmueli, Tree queries: a simple class of queries. ACM Trans. on Database Systems 7(1982), 653-677.

[GT] N. Goodman and Y. C. Tay, A characterization of multivalued dependencies equivalent to a join dependency. Infornation Processing Letters 18(1984), 261-266.

[Gr1] M. H. Graham, On the universal relation. Technical Report, Univ. of Toronto, Sept. 1979.

[Gr2] M. H. Graham, Path expressions in databases. Proc. 2nd ACM SIGACT-SIGMOD Symp. on Principles of Database Systems, 1983, Atlanta, 366-378.

[Gr3] M. H. Graham, Functions in databases. ACM Trans. on Database Systems 8(1983), 81-109.

[GM] M. H. Graham and A. O. Mendelzon, Notions of dependency satisfaction. Proc. 1st ACM SIGACT-SIGMOD Symp. on Principles of Database Systems, 1983, Los Angeles, 177-188.

[GMV] M. H. Graham, A. O. Mendelzon, and M. Y. Vardi, Notions of dependency satisfaction. Stanford University Technical Report STAN-CS-83-979, Aug. 1983. To appear in J. ACM.

[GY] M. H. Graham and M. Yannakakis, Independent database schemes. J. Computer and System Sciences 28(1984), 121-141.

[GJ] J. Grant and B. E. Jacobs, On the family of generalized dependency constraints. J. ACM 29(1982).

[GL] Y. Gurevich and H. R. Lewis, The inference problem for template dependencies. Proc. First ACM SIGACT-SIGMOD Principles of Database Systems (1982), Los Angeles, 221-229.

[HITK] K. Hagihara, M. Ito, K. Taniguchi, and T. Kasami, Decision problems for multivalued dependencies in relational databases. SIAM J. Computing 8(1979), 247-264.

[He] I. J. Heath, Unacceptable file operations in a relational data base. Proc. 1971 ACM-SIGFIDET Workshop on Data Description, Access, and Control, 1971, San Diego.

[Ho] P. Honeyman, Testing satisfaction of functional dependencies. J. ACM 29(1982), 668-677.

[HLY] P. Honeyman, R. E. Ladner, and M. Yannakakis, Testing the universal instance assumption. Information Processing Letters 10(1980), 14-19.

[Hu] R. Hull, Finitely specifiable implicational dependency families. J. ACM 31(1984), 210-226.

[JK] D. S. Johnson and A. Klug, Testing containment of conjunctive queries under functional and inclusion dependencies. J. Computer and System Sciences 28(1984), 167-189.

[KCV] P. C. Kannelakis, S. S. Cosmadakis, and M. Y. Vardi, Unary inclusion dependencies have polynomial-time inference problems. Proc. 15th ACM SIGACT Symp. on Theory of Computing, 1983, Boston, 264-277.

[LMG] K. Laver, A. O. Mendelzon, and M. H. Graham, Functional dependencies on cyclic database schemes. Proc. ACM SIGMOD Symp. on Management of Data, 1983, San Jose, 79-91.

[Le] H. Lewis, Complexity results for classes of quantificational formulas. J. Computer and Systems Sciences 21(1980), 317-353.

[Li] Y. E. Lien, On the equivalence of database models. J. ACM 29(1982), 333-363.

[LD] L. Liu and A. Demers, An algorithm for testing lossless join property in relational databases. Information Processing Letters 11(1980), 73-76.

[Ma] D. Maier, The Theory of Relational Databases. Computer Science Press, Rockville, Maryland, 1983.

[MMSU] D. Maier, A. O. Mendelzon, F. Sadri, and J. D. Ullman, Adequacy of decompositions of relational databases. J. Computer and Systems Sciences 21(1980), 368-379.

[MMS] D. Maier, A. Mendelzon, and Y. Sagiv, Testing implications of data dependencies. ACM Trans. on Database Systems 4(1979), 455-469.

[MRW] D. Maier, D. Rozenshtein, and D. S. Warren, Windows on the world. Proc. ACM SIGMOD Symp. on Management of Data, 1983, San Jose, 68-78.

[MSY] D. Maier, Y. Sagiv, and M. Yannakakis, On the complexity of testing implications of functional and join dependencies. J. ACM 28(1981), 680-695.

[MU] D. Maier and J. D. Ullman, Connections in acyclic hypergraphs. Proc. 1st ACM SIGACT-SIGMOD Symp. on Principles of Database Systems (1982), Los Angeles, 34-39.

[MUV] D. Maier, J. D. Ullman, and M. Y. Vardi, On the foundations of the universal relation model. ACM Trans. on Database Systems 9(1984), 283-308.

[Me] A. Mendelzon, Database states and their tablueax. ACM Trans. on Database Systems 9(1984), 264-282.

[MM] Mendelzon, A. O. and D. Maier, Generalized mutual dependencies and the decomposition of database relations. Proc. Int. Conf. on Very Large Data Bases, (A. L. Furtado and H. L. Morgan, eds.), 1979, 75-82.

[Mi1] J. C. Mitchell, Inference rules for functional and inclusion dependencies. Proc. 2nd ACM SIGACT-SIGMOD Symp. on Principles of Database Systems, 1983, Atlanta, 58-69.

[Mi2] J. C. Mitchell, The implication problem for functional and inclusion dependencies. Information and Control 56(1983), 154-173.

[Ni] J-M. Nicolas, First order logic formalization for functional, multivalued, and mutual dependencies. Proc. ACM SIGMOD Symp. on Management of Data, 1978, 40-46.

[Pa] J. Paredaens, Transitive dependencies in a database scheme. MBLE Research Report R387, 1979.

[PJ] J. Paredaens and D. Janssens, Decompositions of relations: a comprehensive approach. In Advances in Data Base Theory - Vol. 1 (H. Gallaire, J. Minker, and J-M. Nicolas, eds.), Plenum Press, 1981, 73-100.

[PP] D. S. Parker and K. Parsaye-Ghomi, Inference involving embedded multivalued dependencies and transitive dependencies. Proc. ACM SIGMOD Symp. on Management of Data, 1980, 52-57.

[Ri1] J. Rissanen, Independent components of relations, ACM Trans. on Database Systems 2(1977), 317-325.

[Ri2] J. Rissanen, Theory of relations for databases - a tutorial survey. Proc. 7th Symp. on Math. Found. of Comp. Science, 1978, Lecture Notes in Computer Science - Vol. 64, Springer-Verlag, 537-551.

[Ri3] J. Rissanen, On equivalence of database schemes. Proc. 1st ACM SIGACT-SIGMOD Symp. on Principles of Database Systems, 1982, Los Angeles, 23-26.

[Ro] H. Rogers, Theory of Recursive Functions and Effective Computability. McGraw-Hill, 1967.

[Sac] D. Sacca, On the recognition of coverings of acyclic database hypergraphs. Proc. 2nd ACM SIGACT-SIGMOD Symp. on Principles of Database Systems, 1983, Atlanta, 297-304.

[SU] F. Sadri and J. D. Ullman, Template dependencies: A large class of dependencies in relational databases and their complete axiomatization. J. ACM 29(1981), 363-372.

[Sag1] Y. Sagiv, An algorithm for inferring multivalued dependencies with an application to propositional logic. J. ACM 27(1980), 250-262.

[Sag2] Y. Sagiv, Can we use the universal instance assumption without using nulls? Proc. ACM SIGMOD Symp. on Management of Data, 1981, 108-120.

[Sag3] Y. Sagiv, A characterization of globally consistent databases and their correct access paths. ACM Trans. on Database Systems 8(1983), 266-286.

[SW] Y. Sagiv and S. Walecka, Subset dependencies and a completeness result for a subclass of embedded multivalued dependencies. J. ACM 29(1982), 103-117.

[SS] M. Schkolnick and P. Sorenson, The effects of denormalization on database performance. The Australian Computer Journal 14(1982), 12-18.

[Sc1] E. Sciore, Real-world MVDs. Proc. ACM SIGMOD Symp. on Management of Data, 1981, 121-132.

[Sc2] E. Sciore, Inclusion dependencies and the universal instance. Proc. 2nd ACM SIGACT-SIGMOD Symp. on Principles of Database Systems, 1983, Atlanta, 48-57.

[Sc3] E. Sciore, A complete axiomatization of full join dependencies. J. ACM 29(1982), 373-393.

[TY] R. E. Tarjan and Yannakakis, M., Simple linear-time algorithms to test chordality of graphs, test acyclicity of hypergraphs, and selectively reduce acyclic hypergraphs, SIAM J. Computing 13(1984), 566-579.

[Ul] J. D. Ullman, Principles of Database Systems. Computer Science Press, Rockville, Maryland (1982)

[Va1] M. Y. Vardi, The implication problem for data dependencies in relational databases. Ph.D. Dissertation (in Hebrew), The Hebrew University in Jerusalem, Sept. 1981.

[Va2] M. Y. Vardi, The implication and finite implication problems for typed template dependencies. J. Computer and System Sciences 28(1984), 3-28.

[Va3] M. Y. Vardi, On decomposition of relational databases. Proc. 23rd IEEE Symp. on Foundation of Computer Science, Chicago, 1982, 176-185.

[Va4] M. Y. Vardi, Inferring multivalued dependencies from functional and join dependencies. Acta Informatica, 19(1983), 305-324.

[Va5] M. Y. Vardi, A note on lossless database decompositions. Information Processing Letters 18(1984), 257-260.

[Ya] M. Yannakakis, Algorithms for acyclic database schemes. Proc. Int. Conf. on Very Large Data Bases, 1981, 82-94.

[YP] M. Yannakakis and C. Papadimitriou, Algebraic dependencies. J. Computer and System Sciences 25(1982), 3-41.

[YO] C. T. Yu and M. Z. Ozsoyoglu, An algorithm for tree-query membership of a distributed query. Proc. IEEE COMPSAC, 1979, 306-312.

[Za] C. Zaniolo, Analysis and design of relational schemata for database systems, Ph.D. Dissertation, Tech. Rep. UCLA-ENG-7669, UCLA, July 1976.

IBM Research Laboratory, San Jose, California 95193

EMP	DEPT	MGR
Hilbert	Math	Gauss
Pythagoras	Math	Gauss
Turing	Computer Science	von Neumann

Figure 2.1

EMP	DEPT	MGR
Hilbert	Math	Gauss
Pythagoras	Math	Gauss
Turing	Computer Science	von Neumann
Cauchy	Math	Euler

Figure 2.2

EMP	DEPT
Hilbert	Math
Pythagoras	Math
Turing	Computer Science

DEPT	MGR
Math	Gauss
Computer Science	von Neumann

Figure 2.3

STORE	ITEM	PRICE
Macy's	Toaster	$20.00
Sears	Toaster	$15.00
Macy's	Pencil	$ 0.10

Figure 2.4

STORE	ITEM
Macy's	Toaster
Sears	Toaster
Macy's	Pencil

ITEM	PRICE
Toaster	$20.00
Toaster	$15.00
Pencil	$ 0.10

Figure 2.5

EMP	SALARY	CHILD
Hilbert	$80K	Hilda
Pythagoras	$30K	Peter
Pythagoras	$30K	Paul
Turing	$70K	Tom

Figure 2.6

EMP	CHILD	SKILL
Hilbert	Hilda	Math
Hilbert	Hilda	Physics
Pythagoras	Peter	Math
Pythagoras	Paul	Math
Pythagoras	Peter	Philosophy
Pythagoras	Paul	Philosophy
Turing	Tom	Computer Science

Figure 2.7

SUPPLIER	PROJECT	DATE

SUPPLIER	PART	COST

SUPPLIER	PART	PROJECT

Figure 5.1

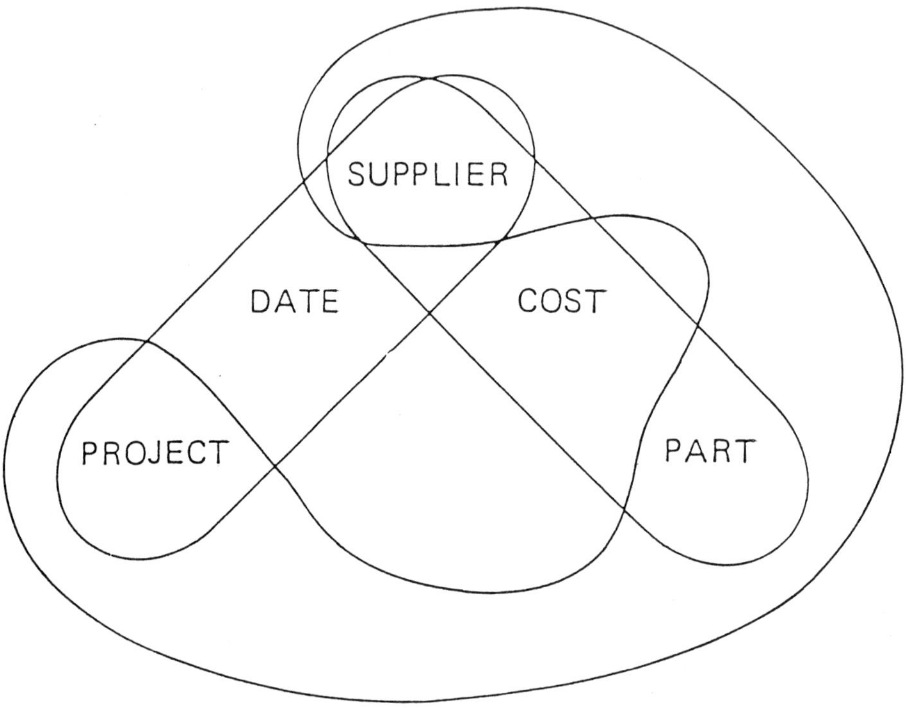

Figure 5.2

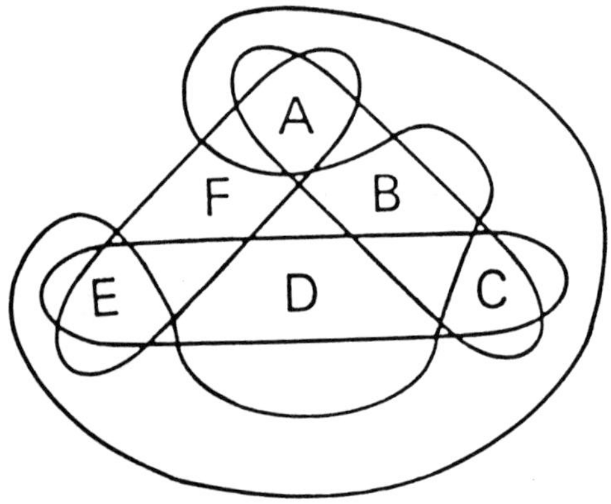

Figure 5.3

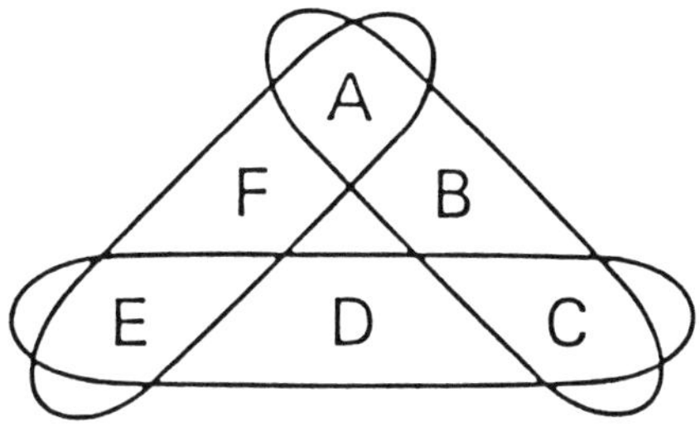

Figure 5.4

A	B
0	0
1	1

B	C
0	0
1	1

A	C
0	1
1	0

Figure 5.5

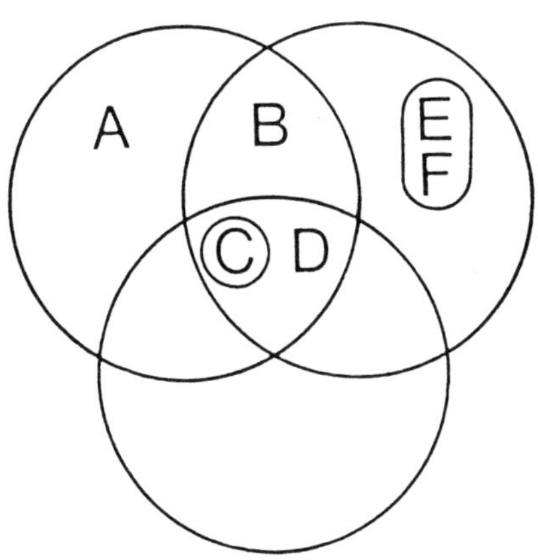

Figure 5.6

Proceedings of Symposia in Applied Mathematics
Volume 34, 1986

TRANSACTION MANAGEMENT

Hector Garcia-Molina

ABSTRACT. Database management systems must guarantee that user programs, called transactions, are executed as atomic units. This means that partial results from a transaction should not be observable by other transactions, even if transactions are executed concurrently and if failures occur. The mechanisms for ensuring this are surveyed and compared.

1. Introduction. One of the most important responsibilities of a database management system is to protect its data from invalid modifications or deletions. It is clearly not desirable to have a system that reports the age of an employee as a negative number, or makes a reservation on a flight that does not exist.

To stipulate what "invalid data" means, it is common to define *integrity or consistency constraints* on the data. These are predicates that valid data must satisfy. For instance, we may state that the age of an employee must be a positive integer less than 100, or that there should not be two employees with the same social security number. The constraints could also be more complex, e.g., they could state that no employee should make twice the average salary of all employees, or that when a salary is modified, it should not be decremented.

What causes these constraints to be violated? There are many causes, but most of them can be grouped into the following four categories:

(a) *Input Data Error.* The integrity constraints may not hold in a database because the data input by the users is incorrect. For example, someone may have actually typed the employee's age as a negative number.

(b) *Program Bug.* It is also possible that the programs, either those written by the users or those provided by the system itself, malfunction and write incorrect data into the database. For instance, the system may have computed a negative balance in an account, and then stored, incorrectly, that value in the age field of an employee.

(c) *Data Sharing.* If users concurrently modify the database, inconsistencies may arise. Say two users want to reserve the last seat on an airline flight. If no precautions are taken, the programs for both users could decide that there are enough seats, and both programs could write in the database that they have the last seat.

0160-7634/86 $1.00 + $.25 per page

(d) *Failures.* A hardware failure or malfunction can also lead to the infraction of an integrity constraint. For example, consider a program that is transferring funds from one account into another. If the computer crashes after the funds are withdrawn but before they are deposited, an incorrect state (with missing money) will exist.

In these notes we concentrate on the last two categories and present mechanisms for preventing or detecting integrity violations. The key idea is to have the system execute each user program or request as if it were an *atomic action.* If each program is indivisible, then different users cannot interfere with each other. (If programs are executed as atomic units, only one of the users requesting the last seat can get it. The first program to read the available number of seats will have to complete before the other reservation program can run.) Furthermore, if the system manages to execute programs atomically, then some (but not all) of the problems due to hardware failures are solved because programs can no longer remain half done.

The atomic action concept is especially attractive because it does not require knowledge of the consistency constraints. That is, a mechanism that ensures atomic operations is independent of the database application, and hence, the users are not burdened with the definition of the constraints. On the other hand, the only way to protect against input data errors and programming bugs is by explicitly defining the constraints so that the system can check and enforce them. As one might suspect, the mechanisms for handling these last types of infractions are different from the ones for data sharing and failure violations, and are hence not covered in these notes.

There are two reasons why the implementation of atomic actions is a particularly challenging problem:

- **Efficiency.** Any solution is not satisfactory, only an efficient one is. For instance, it is not acceptable to execute *all* programs serially, one after the other, in order to ensure that programs or user requests are atomic. (If two programs do not interfere, there is no reason why they cannot be executed concurrently, without delays.) Furthermore, in database systems, accesses to the disk are very expensive and must be minimized. Thus, the difference between a mechanism that makes two disk requests for an operation and one that makes only one can be very significant. (Those readers familiar with analysis of algorithms will shortly see that all the algorithms we present here are linear in the number of database operations made by the programs. We clearly can do no better than this. However, its the constants of proportionality that make all the difference between strategies.)
- **Failures.** There are a large number of types of hardware failures (ranging from a parity error in memory to a fire in the computer room), and it is impractical to attempt to protect against all within the database management system. Moreover, even for a small number of failures, there are a large number of cases that must be considered in the analysis of the algorithms. Thus, the challenge is to precisely

define the most common failures in a database environment, and to show (or at least give convincing arguments) that the failure protection mechanisms actually work.

To survey existing mechanisms that implement the atomic action concept, we start by addressing, in Section 2, the data sharing problem in an environment where failures do not occur. This assumption simplifies the presentation, but in Section 3 we relax it and study some common failures in a database system and what steps can be taken for protection.

In our discussion we concentrate on the principal concepts and will often ignore implementation details. The reader should, however, keep in mind that the actual concurrency control and crash recovery mechanisms found in database systems are often considerably more complex than what is presented here.

2. Concurrency control. We can model the database as a collection of *objects*, where each object has a name and a value. From our point of view, it is not important what these objects are: they can be the records in the database, or the disk blocks that hold the database, or some other unit into which the database is partitioned. We say that the database is consistent (or in a consistent state) if the values of the objects satisfy all of the consistency constraints.

We can model the user programs or requests as collections of *actions*. Each action is an indivisible operation that *may* modify the database. There are, of course, many types of actions; some only read the database, others write it, others increment a value. However, for the time being let us only consider a single type of action that encompasses all types of operations. We assume that these actions are atomic. That is, we assume that there is some low level database mechanism that excludes concurrent actions on the same object. (This is a reasonable assumption as long as the object is smaller than a disk block.) As mentioned earlier, in this section we ignore hardware failures.

Even if actions are atomic, the consistency constraints cannot be enforced at all times. To illustrate, consider a constraint that specifies that the value of object A must equal that of B (i.e., $A = B$), and say that A and B are integers. If a user wishes to increment both values by 100, his program must execute at least two actions, and between these, the constraint will not hold. However, the actions of the user program, if considered as a unit, will preserve consistency. That is, if initially $A = B$, after the program that increments them both by 100 runs, A will again equal B.

Thus, we assume that the user programs consist of one or more *transactions*, where each transaction consists of a series of actions. Each transaction will transform any consistent database state into another consistent state, as long as it is executed without interference form other transactions. However, if the actions of transactions are interleaved, then consistency may be compromised. The following example illustrates this.

Consider once again the constraint $A = B$ and the two transactions

T_1: $x \leftarrow A$	T_2: $y \leftarrow A$
$x \leftarrow x+100$	$y \leftarrow 2y$
$A \leftarrow x$	$A \leftarrow y$
$x \leftarrow B$	$y \leftarrow B$
$x \leftarrow x+100$	$y \leftarrow 2y$
$B \leftarrow x$	$B \leftarrow y$

where x and y are local variables in the transactions. (Each line above represents one action. Some actions, like $x \leftarrow 2x$, do not access the database.) The two transactions are valid because they each preserve the consistency constraints. Now suppose that the system executes these actions in the following order:

Execution E_1:

(T_1) $x \leftarrow A$
(T_1) $x \leftarrow x+100$
(T_1) $A \leftarrow x$
(T_2) $y \leftarrow A$
(T_2) $y \leftarrow 2y$
(T_2) $A \leftarrow y$
(T_2) $y \leftarrow B$
(T_2) $y \leftarrow 2y$
(T_2) $B \leftarrow y$
(T_1) $x \leftarrow B$
(T_1) $x \leftarrow x+100$
(T_1) $B \leftarrow x$

The system may want to execute the actions in this order for efficiency. For example, it may be more efficient to perform all the actions on object A (for both T_1 and T_2) before moving the disk head to the location of object B. However, it easy to see that this execution may violate consistency. For instance, if initially $A = B = 0$, then after the execution A will have a value of 200, while B will contain 100. Thus, the users would not be satisfied with a system that could execute this sequence of actions.

Observe that there are other interleavings of T_1 and T_2 that never violate consistency. For instance, the executions

Execution E_2:

(T_1) $x \leftarrow A$
(T_1) $x \leftarrow x+100$
(T_1) $A \leftarrow x$
(T_2) $y \leftarrow A$
(T_2) $y \leftarrow 2y$
(T_2) $A \leftarrow y$
(T_1) $x \leftarrow B$
(T_1) $x \leftarrow x+100$
(T_1) $B \leftarrow x$
(T_2) $y \leftarrow B$
(T_2) $y \leftarrow 2y$
(T_2) $B \leftarrow y$

and

Execution E_3:

(T_1) $x \leftarrow A$
(T_1) $x \leftarrow x+100$
(T_1) $A \leftarrow x$
(T_1) $x \leftarrow B$
(T_1) $x \leftarrow x+100$
(T_1) $B \leftarrow x$
(T_2) $y \leftarrow A$
(T_2) $y \leftarrow 2y$
(T_2) $A \leftarrow y$
(T_2) $y \leftarrow B$
(T_2) $y \leftarrow 2y$
(T_2) $B \leftarrow y$

will always produce a consistent state, as long as $A = B$ initially.

Why is it that the last two executions have this property, and the first one does not? It is easy to see that E_3 preserves consistency because it is *serial*, i.e., T_2 starts once T_1 completes. If $A = B$ initially, transaction T_1 guarantees the constraint holds when it completes. And since T_2 is starting with a consistent state, it too guarantees that A will equal B when it completes.

Execution E_2 also preserves consistency because it is *equivalent* to the serial one. That is, in that second execution we could move the last three actions of T_1 back in time without affecting the values read or written by the transactions. (These actions affect B, while the first three actions of T_2 affect A.)

On the other hand, there is no way that execution E_1 could be converted into a serial one without affecting the values read or written by the transactions. Specifically, if we look at the third ((T_1) $A \leftarrow x$) and the fourth actions ((T_2) $y \leftarrow A$), we see that their order cannot be interchanged without changing what T_2 reads. Therefore, we see that in any rearrangement of this execution into an equivalent serial one, T_1 would have to precede T_2. In a similar fashion, by looking at the ninth ((T_2) $B \leftarrow y$) and the tenth actions ((T_2) $x \leftarrow B$), we deduce that T_2 must precede T_1 if we do not wish to change the value read by T_1. Both of these conditions cannot be true, so there is no serial execution that is equivalent.

It is important to note that the observations we have just made in this example are not dependent on the semantics of the actions nor on the constraint we used. In other words, if we write the first execution as

Execution E_1:	(T_1)	action on A
	(T_1)	action on A
	(T_2)	action on A
	(T_2)	action on A
	(T_2)	action on B
	(T_2)	action on B
	(T_1)	action on B
	(T_1)	action on B

we can still deduce that this execution is not equivalent to a serial one, and hence *may* violate consistency. This means that it is not necessary to understand the application and the constraints in order to identify such executions. This in turn means that it is possible to have a general purpose database system that excludes such executions.

We also should point out that there could be executions that are not equivalent to serial ones, and yet preserve consistency in a given application. But to identify these, one must have knowledge of the application. Hence, for this discussion, we will assume that only executions that are equivalent to serial ones are acceptable.

To state our findings more precisely, let us define a *transaction* T_i to be a series of actions $a_{i,1}$, ..., $a_{i,n}$. Each action $a_{i,j}$ acts upon a single database object $o(a_{i,j})$. The order in which the actions of a set of transactions $\mathbf{T}$ are executed by the system is represented by an *execution* E (sometimes called a *schedule*). The execution is also a sequence of actions. It contains all of the actions of the individual transactions (and no others). The actions of each transaction appear in E in the same relative order. That is, if $a_{i,j}$ precedes $a_{i,k}$ in T_i (i.e., $j < k$), then $a_{i,j}$ must precede $a_{i,k}$ in E. An execution is *serial* if the actions of the transactions are not interleaved (i.e., if all actions in E between $a_{i,j}$ and $a_{i,k}$ belong to T_i).

To check if an execution is equivalent to a serial one, we construct its *precedence graph.* The nodes are the transactions in $\mathbf{T}$. There is an arc $T_i \rightarrow T_j$ in the graph if and only if there are actions $a_{i,k}$ and $a_{j,l}$ in E such that $o(a_{i,k}) = o(a_{j,l})$ and $a_{i,k}$ precedes $a_{j,l}$. If arc $T_i \rightarrow T_j$ exists in the precedence graph of E, this means that to rearrange the actions of E into a serial execution, one must place T_i ahead of T_j.

We say that two executions E_1 and E_2 for a set of transactions **T** are *equivalent* if each action $a_{i,j}$ in E_1 has the same effect (e.g., reads and writes the same database values) as its corresponding action in E_2. From this definition it is easy to see that if two executions are equivalent, then their precedence graphs must be identical. An execution that is equivalent to a serial one is called *serializable*. (If the execution is serializable, then each transaction is executed as if were an atomic action.)

The next theorem provides the basis for checking if an execution is serializable.

THEOREM 2.1. *An execution E_1 is serializable if and only if its precedence graph has no cycles.*

Informal Argument. If E_1 is serializable, then there is a serial execution E_S equivalent to E_1. Since the precedence graph for both executions are equal, and since the graph for E_S clearly has no cycles, we conclude that the graph for E_1 is also acyclic.

To show the converse, we start with execution E_1 with an acyclic precedence graph and construct and equivalent serial schedule. Since the graph is acyclic, there is one transaction T_i with no incoming arcs. The actions of this transaction can be moved to the head of E_1 without changing the effect of the execution (i.e., the rest of the transactions in no way affect what T_i reads or writes). Of the remaining transactions, there will again be one without incoming arcs, and we can move this one to the head, following T_i. We proceed in this fashion until E_1 is converted into an equivalent serial execution. ○

2.1 Locking. Theorem 2.1 tells us what executions are serializable and hence acceptable for the users. However, what the system needs is a mechanism for *preventing* non-serializable schedules from occurring in practice. The most common mechanism for this is *locking*. The key idea is that before a transaction accesses object A, it executes a special *lock* action on A. If the system grants the lock, the transaction has reserved A for its exclusive use, and subsequent lock requests by other transactions will be denied. When the transaction decides it no longer needs to access A, it executes an *unlock* action on A. This releases the lock and allows other waiting transactions to gain access to A.

If locking is to work, the transactions must follow the rules: lock an object before executing an action on it, eventually release all locks, and do not lock an object on which you already hold a lock. We call transactions that follow these rules *well formed*. Similarly, the database system must enforce the following rule: no two transactions can have the same object locked at the same time. We call the executions produced by a system that follows this rule *legal*.

When a transaction is denied a lock, it waits until the lock is released. It is possible for a circular wait condition to arise, and in this case the *deadlock* must be broken by aborting at least one of the transactions involved. To abort a transaction, all of the database modifications made by it are undone, and all of the locks held by it are released.

At a later point in time, the transaction is restarted from scratch. In Section 3 we will return to this issue and discuss some of the techniques for undoing the changes made by a transaction.

At least intuitively, it seems that with these locking rules all destructive interference between transactions will be eliminated, and that the resulting executions will all be serializable. However, this is not the case, as the following execution illustrates:

Execution E_4:

(T_1)	lock A
(T_1)	action on A
(T_1)	unlock A
(T_2)	lock A
(T_2)	action on A
(T_2)	lock B
(T_2)	action on B
(T_2)	unlock A
(T_2)	unlock B
(T_1)	lock B
(T_1)	action on B
(T_1)	unlock B

The two transactions, T_1 and T_2, are well formed and the execution is legal, yet arcs $T_1 \rightarrow T_2$ and $T_2 \rightarrow T_1$ are in the precedence graph and the execution is not serializable (by Theorem 2.1).

The problem can be avoided if we demand that transactions do not request additional locks after they have released their first lock. Transactions that obey this extra rule are called *two phase* because they proceed in two steps or phases: in the first they only request locks, and in the second they can only release locks. We call the point where a transaction releases its first lock its *crest*.

In the example above, we can see how the two phase rule makes the execution serializable. Before T_1 releases its lock on A, it must now request its lock on B. This means that T_2 cannot access B until after T_1 has. In turn, this eliminates the arc $T_2 \rightarrow T_1$ from the precedence graph.

THEOREM 2.2. *If each transaction in the set* $\mathbf{T}$ *is well formed and two phase, then any legal execution of* $\mathbf{T}$ *is serializable.*

Informal Argument. Let E be any legal execution, and consider an arc $T_i \rightarrow T_j$ in its precedence graph. If the arc exists, there must be actions $a_{i,k}$ and $a_{j,l}$ in E where $a_{i,k}$ precedes $a_{j,l}$. At some point in between these actions (or possibly at $a_{i,k}$) T_i must have released its lock on $o(a_{i,k})$. At some later point, but before or at $a_{j,l}$, T_j must have requested a lock for this object. (Recall that $o(a_{i,k}) = o(a_{j,l})$.) Therefore, the crest of T_i must precede the crest of T_j in E, since the transactions are two phase.

With this observation, it is easy to see that there can be no cycle $T_1 \rightarrow T_2 \rightarrow \ldots T_m$ in the graph of E. This would mean that the crest of T_1 precedes itself. Therefore, E is serializable. ○

2.2 Other concurrency control mechanisms. We have presented some of the fundamental concepts in database concurrency control. A large number of extensions and variations on these ideas have been proposed, and in this section we briefly survey some of them.

In our discussion we assumed that there was a single type of action. If instead we classify the actions into their different types, it is possible to reduce the number of arcs in the precedence graphs, and this makes more executions serializable. For example, if in execution E_4 given earlier we know that all of the actions of T_1 and T_2 on object A only read it and do not modify it, then we see that the arc $T_1 \rightarrow T_2$ disappears (i.e., the actions on A can be interchanged without affecting the values read or written). Thus, E_4 is now serializable. Having more serializable executions is, of course, desirable because the database system has greater flexibility in scheduling the transactions.

In practice, the most common way to classify actions is into *read* and *read-write* actions. Read actions do not modify the database object referenced, and read-write actions may read and modify the object. Theorem 2.1 still hold, except that now two read actions do not create an arc in the precedence graph of an execution.

The locking mechanism may also be modified to take advantage of the knowledge that some actions only read. A new type of lock, a *shared* or *read lock* is introduced. If a transaction is only going to read an object, it requests a shared lock. Other reading transactions will be granted concurrent shared locks on the same object. Before a transaction executes a read-write action, it requests an *exclusive lock*, which is identical to the locks discussed in the previous section. This locking mechanism leads to higher concurrency, and can be shown to produce serializable executions. (Transactions must still be well formed and two phase.)

A classification of actions into read and write ones (where the write action does not read the object) has also been explored. However, for reasons that are beyond the scope of these notes, there are no practical concurrency control mechanisms that can take advantage of this classification. Thus, real systems usually confine themselves to shared and exclusive locks.

Another question that has received attention is the size of the objects that are locked by the system. As the database is partitioned into finer and finer objects, the allowable concurrency increases. Unfortunately, the number of locks that must be requested (and released) also grows, and this costs in terms of system performance. Hence, for each application there is an optimal object size.

In some cases it is best to form a hierarchy of the database objects, so that they can be locked at different granularities. For instance, we may divide the database into files, and each file into records. If a transaction wishes to access an entire file, it locks the file. However, if it wishes to access a single record, it first checks if the corresponding file is not locked, and then locks the record it wants. While the record is locked, the file is locked with a special warning lock. This indicates to other transactions that some of the records in the file are locked. A transaction wanting to lock the entire file must be delayed when it encounters the warning lock, but other transaction that also want to lock at the record level may proceed. (Any conflicts will be detected when they lock the records they need.)

Finally, not all concurrency control mechanisms utilize locking. *Optimistic* mechanisms let all transactions run with no delays. However, the system keeps track of the objects read and written by each transaction, and when they complete, it checks if the resulting execution would be serializable. Any transaction that (due to the order in which it accessed data) may cause the execution to be non-serializable is aborted. Other mechanisms are based on *timestamps*. Here the system attaches a timestamp to each object indicating the time when the object was last read or written. This information is then used to abort transactions that may cause non-serializable executions.

3. Coping with failures. So far we have assumed that the hardware that executes the actions on behalf of the transactions does not fail. In this section we look at some of the precautions that the database system can take to insure that transactions are still executed as atomic actions. We start by considering a simple failure model, and later will consider other types of failures.

We view the system hardware as consisting of three components: the *CPU*, the *main memory*, and the *secondary storage* (or *disk* for short) where the database itself is held. We assume that the CPU and memory can only have one type of failure: a *crash*. When the CPU crashes, it simply halts and the contents of main memory are lost. At no point does the CPU deviate from its algorithms; it only suspends their execution when a crash occurs. At some undetermined time, the CPU is restarted at a fixed section of code. This is the *crash recovery code* that specifies the actions that must be taken before normal processing is resumed.

For the time being, we assume that the disk is failure free. If the CPU crashes while an action that modifies the disk was in progress, the action will either complete successfully, or it will not be performed at all. In no case will the value of the object being modified be ruined or lost. Similarly, we assume that the disk will not spontaneously lose data stored in it.

When the system crashes, transactions that were in progress are left incomplete. This means that the database may be in an inconsistent state. (Recall that a transaction only preserves consistency if all of its actions are executed.) Therefore, when the system recovers, it has only two choices: either the transactions in progress are aborted or the missing actions are executed.

Let us study the first possibility. To abort a transaction, we must restore the values of the objects that were modified to the values they had when the transaction started. For this, the system must save the old values. They are saved in what is called an *undo log* or *journal*. There is a single log for the entire system where all the old values are stored.

Since main memory is lost during a crash, the undo log must be kept on disk. The log is only read after crashes, so its organization should be optimized for fast writes. Typically, the old values (with an identification of the transaction involved) are appended to the end of a file. Furthermore, the old value must be written safely on the disk *before* the database is actually modified. Otherwise, a crash between the time the database is written and the old value is stored on disk can get us into trouble. By writing the old value to the log first, the worst that can happen is that we restore an "old" value to an object that was never modified. The technique of writing the log first is referred to as the *log write ahead protocol*.

To illustrate the operation of the log, consider a transaction T_1 that is to execute three actions that modify the database: $a_{1,1}$, $a_{1,2}$, $a_{1,3}$. When the system initiates T_1, it makes a begin-transaction entry in the log. When T_1 issues action $a_{1,1}$, the system obtains the old value of $o(a_{1,1})$. It stores this value, together with an identification for T_1 and the name of object $o(a_{1,1})$ in the log. (There could be entries for other transactions in the log between the begin-transaction entry for T_1 and this last entry.) Once this is done, the new value for $o(a_{1,1})$ is written into the database. The other actions are processed in a similar fashion. When T_1 completes, it informs the system, and an end-transaction entry for T_1 is made in the log.

Suppose that the CPU crashes before $a_{1,3}$ is executed. When the system recovers, it scans the log in reverse order. When it encounters an entry for T_1 with an old value, it realizes that T_1 was in progress at the time of the crash and must be aborted. The old value is stored in $o(a_{1,2})$, and the scan continues. Later, the entry for $o(a_{1,1})$ is encountered and processed. When the begin-transaction entry for T_1 is reached, the system knows that T_1 has been successfully aborted.

Two other failure scenarios are worth considering. Suppose that the CPU crashes during the recovery described in the previous paragraph, just after the old value for $o(a_{1,2})$ was restored. In this case, the second recovery proceeds just like the first. When the system finds the entry for $o(a_{1,2})$ in the log, it will reinstate the old value once again, but this does no harm. Later, the value for $o(a_{1,1})$ will be restored as described above.

Next, suppose that the CPU crashes after T_1 completes. In this case, when the log is scanned at recovery time, the end-transaction entry for T_1 will be encountered first. Thus, the system will know that T_1 completed and need not be aborted.

As we have described the undo log, upon recovery it must be scanned in its entirety in order to discover all the transactions that must be aborted. This is not efficient because the old values for the pending transactions will usually be found at the very end of the log. This full scan can be eliminated in the system periodically performs *checkpoints*. At a checkpoint, the system writes a special log entry that identifies all the transactions that are currently in progress. At recovery time, the system proceeds as described earlier until it reaches a checkpoint record. At that time, the pending transactions to be aborted are identified, and the log need only be scanned until all of their begin-transaction entries have been located.

The undo log we have described is not only useful for crash recovery; it can also be used to abort transactions involved in deadlocks or transactions that voluntarily wish to abort (e.g., a user is reserving seats on some flights, but then decides to cancel his trip). To abort a single transaction, the system proceeds as described earlier, except, of course, that only the entries corresponding to the aborted transaction are considered.

Our last, but very important comment on undo logs, has to do with the interaction of concurrency control and crash recovery (and transaction abort). Consider a transaction T_1 that has not completed but has already executed action $a_{1,j}$ to modify a database object. If T_1 releases it lock on $o(a_{1,j})$, another transaction T_2 can read the value of $o(a_{1,j})$. Yet, since T_1 has not completed, it could be aborted at any time (due to a crash or deadlock), rendering the value stored in $o(a_{1,j})$ and being used by T_2 invalid.

There are two strategies for solving this problem. The first is for the system to keep track of these occurrences, and when a transaction like T_1 is aborted, it will also abort all transactions (like T_2) that read data produced by T_1. Naturally, any transactions that read data written by T_2 must also be aborted, so this process is known as *cascading* aborts. Most real database systems do not follow this strategy because abortions are expensive operations.

The second strategy is to delay the release of locks until it is certain that a transaction will not be aborted. This time is called the *commit* point of a transaction. With an undo log, the commit point of a transaction occurs when the transaction completes its last action, and hence *all* locks are released at this time. This lock release strategy of course satisfies the two phase locking requirement discussed in Section 2.1.

So far, we have only considered one of the two approaches for returning the database to a consistent state after a crash: aborting or undoing pending transactions. The second approach is to complete the pending transactions. The problem here is that the state of the pending transactions (e.g., their local variables), which is necessary to restart them, was stored in main memory and has been lost in the crash. Therefore, the only solution is to have the transaction generate the new values that will go in the database before the crash occurs.

In other words, before the transaction writes anything in the database, it must safely store on disk all of the values it will write. These *new values* are stored in a *redo log* or *journal*, which is very similar to the undo log. When a transaction starts, it writes a begin-transaction entry in the log. When it wishes to modify an object, it writes an entry in the log giving the new value and the name of the object and transaction involved. The actual write into the database, however, is postponed. When the transaction decides it needs to perform no more actions, an end-transaction entry is made in the log.

At this time the commit point has been reached for this transaction. If a crash occurs before this point, the transaction is aborted. (The abortion is simple since the transaction has not modified the database. Only an abort-transaction entry has to be made in the log to indicate that the redo entries for this transaction can be ignored.) After the commit point, the writes generated by the transaction are actually installed in the database. If a crash occurs here, the transaction will not be aborted. Instead, the log will be consulted and the writes generated by the transaction will be installed in the database. (Some of the new values may already be in place, but writing a new value in the database more than once creates no problems.)

As with the undo log, checkpoints can be made periodically to speed up recovery after a crash. Also, the locks held by a transaction are only released after the commit point. In this case, however, the have to be held until the actual writes to the database complete, and hence they do not have to be released all at once.

Both the undo and redo logs can be used to make the database consistent after a crash. The redo log has an important advantage over the undo log: as we will see later, it can be used for reconstructing the database if the disk fails. On the other hand, the redo log is advantageous because it does not delay the writes to the database until after the commit point. So in practice, many systems use both types of logs, or actually a single combined *undo-redo log* that combines the advantages of both mechanisms. For each write to the database, both the old and the new values are stored in the log. This log gives the database system grater flexibility in scheduling the writes, for they can now occur either before or after the commit point of the transaction.

Logs are not the only mechanisms for making transactions atomic in the face of CPU crashes. A second technique, called *shadow paging*, is less common, but may be useful in small databases. The key idea is to structure the database as a tree. The nodes (or pages) of the tree are never modified directly. If changes have to be made, shadow nodes (or pages) are created. These are copies of the original nodes where the modifications are actually carried out. To modify the database itself, a new root node is created in memory that points to the new version of the tree. This new version consists of the nodes of the original tree that were not modified, plus the new shadow nodes. Finally, the new root in memory is copied over the old root on disk. Assuming that this root occupies a single disk page, this operation can be considered atomic. Thus, if a crash occurs before the new root is written out, all the changes are lost. If a crash occurs after the new root is in place, then all of the modifications made by the transaction will be in the database.

Notice that if a leaf of the tree is modified (and most of the database will be in the leafs), all the nodes in the path from that leaf to the root will also be modified. (For instance, the direct parent of the leaf must be modified so it points to the shadow leaf. The grandparent must now point to the new shadow parent, and so on.) As the database becomes larger, the depth of the tree increases and the number of nodes that must be shadowed also grows. Therefore, the shadow mechanism becomes less attractive as the database grows. Another problem is that the database cannot be stored contiguously and sequentially on the disk. Finally, additional safeguards must be taken if the disk can fail, but this is discussed in the next section.

3.1 Disk failures. A second, common failure that database systems must protect against is a *disk failure* (sometimes called a *media failure*). A disk may fail in two ways. Instead of executing a write command properly, it may write garbage, ruining the object(s) that were being accessed. It may also fail spontaneously and ruin objects that were not being accessed by an action (usually called a head crash). In either case, the database, or the log if that is where the malfunction occurred, is lost.

It is usually assumed that the loss of data from a disk failure can be detected by the database system when it attempts to read the ruined data. This is achieved by adding *error detecting codes* to the object values when they are written out. If at a later point in time the system reads the object and the value it reads is not the same one it stored earlier, the error detecting code will not match, and the failure will have been detected. Of course, it is possible that a failure modifies both the object and its code, so that the new ruined object matches the new ruined code. However, we assume that the probability of such an event is insignificant. (The probability of an undetected failure can be made arbitrarily low by increasing the size of the code associated with an object.)

The basic solution to the disk failure problem is to replicate the data, so that if one copy fails we can still recover the data from the other copy or copies. The copies should be made on different physical disk units, so that the failure of one will not affect the others. To protect against disasters like floods or fires, the copies should also be in different physical locations, if possible.

Independent disk failures could ruin two or more copies of the data, so we must also ensure that between failures there is sufficient time for the system to detect the loss of data and to make new copies of the correct data. Again, we cannot guarantee that simultaneous failures will not destroy all copies. We can, however, make a large enough number of copies (if we can afford them) so that the probability of losing them all before the system can make extra copies is very low. In most real applications, two or three copies are deemed sufficient.

All mechanisms that protect against disk failures follow the general principles we have outlined so far. However, there are a large number of ways in which to implement these ideas, and specifically, there are many techniques for updating the multiple database copies.

The simplest strategy is probably to keep n full copies, and to update them simultaneously as the database is updated. As we discussed earlier, the copies should be on different physical devices, so updating them concurrently is feasible. To read the database, the system can pick any one of the copies since they are all identical. If an error is detected on a read, then the other copies can be read, and the ruined copy can be immediately restored. Similarly, the system should periodically scan all the copies looking for errors, and take corrective action when they are found.

The main drawback of this strategy is its cost. Keeping n full copies on disk is expensive, so a common solution is to keep all but one of the copies on a cheaper medium like magnetic tape. Reading and writing tapes is much slower than accessing disks, so typically the copies on tape are kept *off-line*. That is, there is only one up to date copy of the database, and it resides on disk. The other copies are on tape and are not directly accessible by the system. Periodically (say at night), new up to date copies of the database are made and stored. It is simplest to make these ***backup*** copies (or database ***dumps***) when the database is not being modified, but is is also possible to make copies while the system is in operation.

The backup copies safeguard the database as it existed when the copies were made, but do not protect the current version of the database. In other words, if the main copy of the database is lost and we go back to a copy on tape, we will lose all the updates that were performed between the backup to tape and the disk failure. This is not acceptable in most database applications. To solve this problem, we must also make copies of the new database values as they are generated, at least until new backup copies are made on tape.

The new values can be stored in a redo log, as discussed in the previous section. The procedure for recovering from the loss of the database is then as follows. When a backup copy (or copies) is made, a database-dumped entry is made in the redo log indicating this. When the database is lost, the system loads the lastest backup copy onto the disk. The lastest database-dumped entry is located in the redo log, and from that point the log is scanned in a forward direction. Each new value found is installed in the database. When the end of the log (i.e., the most recent entry) is reached, the database that existed just before the failure has been reconstructed, and transaction processing may resume. (If the database loss was accompanied by a CPU crash, then this recovered state may be inconsistent, and pending transactions will have to be terminated, as discussed earlier.)

The log should be on a separate physical device to isolate it from failures of the database disk. Furthermore, it may be desirable to make several copies of the log itself to protect it from failures. Replicating the log is not as expensive as replicating the database because it is not as large as the database. (Notice that when a backup copy of the database is made, the redo log up to that point may be discarded.) Also, since the log is a file that is written sequentially, it can be kept on tape.

The new values used to reconstruct the database do not necessarily have to be kept in the form of a redo log. For instance, instead of organizing the new object values sequentially according to the time they were created, we can store them by object name. For each database object that was modified since the last backup, the database can keep its latest value in a special data structure, separate from the full database. This structure can be implemented in a variety of ways.

3.2 Other failures. In these notes we have only covered the two most common types of failures encountered in database systems: CPU crashes and disk failures. Most commercial systems only consider these types, but some research ultra-reliable systems are beginning to consider additional ones.

One noteworthy type of failure being studied is a CPU malfunction that causes it to deviate from its algorithms. If a CPU can suddenly execute invalid and arbitrary instructions, the database it controls can easily be destroyed (e.g., when the CPU fails, it may start writing junk into the database). The only solution in this case is to replicate, not only the database, but the CPU itself. Thus, the system may contain three or more identical processors and databases. Each processor executes each transaction on its database, and the outputs of the transaction are compared. The result produced by a majority of the processors is taken to be the correct one.

4. Conclusions. In these notes we have briefly outlined some of the mechanisms used by database systems to protect the integrity of their data. The fundamental concept is that the user requests or transactions should be executed as atomic actions. Neither failures nor concurrent execution of the transactions should make it possible to observe the partial results of a transaction. Moreover, once a transaction completes, its effects should be durable. That is, the data it wrote into the database should not be lost due to a failure of the secondary storage device.

We stress, once again, that the mechanisms presented here are highly simplified. We have glossed over many of the details in order to stress the basic concepts. To illustrate some of the problems and issues we did not touch on, we briefly mention two.

In addition to executing actions on the database, transactions also execute actions that impact the world outside. For example, transactions send messages to users and transactions dispense cash at automatic tellers. The system must treat such actions differently and carefully. If a transaction is aborted, there is no way to get the cash back into the teller machine. Similarly, if we redo the transaction, we should not dispense the cash twice.

Databases are in many cases not located on a single computer, but are *distributed* over a collection of machines that are connected by a communication network. The concept of an atomic transaction is still very important in this environment, but our mechanisms that implement them must be modified. The concurrency control and crash recovery tasks must be performed by all the computers in an integrated and coordinated fashion.

BIBLIOGRAPHY.

These notes are based on and summarize the work of many database researchers. Specifically, Section 2 is based mainly on [2], the first paper to formally discuss the concepts of serializable execution and of two phase locking. Section 3 was strongly influenced by [3], a comprehensive work that presents many of the internals of database systems. The books [1], [4], and [5] present excellent overviews of the database filed in general, including most of the topics covered here.

[1] C. J. Date, *An Introduction to Database Systems, Volume II*, Addison-Wesley, Reading, Massachusetts, 1983.

[2] K. P. Eswaran, J. N. Gray, R. A. Lorie, and I. L. Traiger, The notions of consistency and predicate locks in a database system, *Communications of the ACM*, 19, 11, Nov. 1976, pp. 624-633.

[3] J. N. Gray, Notes on database operating systems, *Operating Systems: An Advanced Course*, R. Bayer, R. M. Graham and G. Seegmuller, editors, Springer-Verlag, 1979, pp. 393-481.

[4] J. D. Ullman, *Principles of Database Systems*, (Second Edition), Computer Science Press, Rockville, Maryland, 1982.

[5] G. Wiederhold, *Database Design*, McGraw-Hill, New York, 1977.

DEPARTMENT OF ELECTRICAL ENGINEERING AND COMPUTER SCIENCE, PRINCETON UNIVERISTY, PRINCETON N.J. 08544

Current Address; Department of Electrical Engineering and Computer Science, Princeton University, Princeton N. J. 08544

Proceedings of Symposia in Applied Mathematics
Volume 34, 1986

FUNDAMENTAL DATABASE ISSUES [1]

Barry E. Jacobs

Department of Computer Science
University of Maryland
College Park, Maryland
and
National Aeronautics and Space
Administration
Goddard Space Flight Center
Greenbelt, Maryland

ABSTRACT

This purpose of this paper is to provide an introduction to database systems technology. The paper first reviews the three basic data models- relational, hierarchical and network through illustrative examples. It next uses these examples to present some fundamental database issues. The use of first-order logic is then seen as a formalism for databases when the data model is relational. The paper concludes by discussing how database logic can serve the relational, hierarchical and network cases as first-order logic serves the relational case.

[1] Barry Jacobs, APPLIED DATA BASE LOGIC 1: FUNDAMENTAL DATA BASE ISSUES, a forthcoming Prentice-Hall, Inc., 1984 publication. Excerpted by permission of the publisher.

0160-7634/86 $1.00 + $.25 per page

1. DATABASES

Roughly speaking, a *database* is a collection of data and operations that represents some aspects of the real world. For example, a university database can be regarded as representing parts of a university environment. Hence, one may regard a database as a "model" of that part of reality that one is interested in. Consequently, when we use a database, questions about the real world are reduced to questions about the model.

There are three major types of databases- relational, hierarchical and network. The differences between them are based on the way in which data in the database appears to the user. Illustrations of the three types are given below. A *database management system* is a large scale computer program that is responsible for supporting many databases. Depending on the type of databases it maintains, a database management system is called relational, hierarchical or network.

Relational Databases

Relational databases are databases in which data appears to the user in one or more separate tables. These tables, called *relations,* have rows which are called *records.* For example, Figure 1 contains an instance of a relational database. The database represents student transcript information at a university. The relation TRANSCRIPT has records that indicate the student id#, student name, semester, year, department, course#, section#, course title, grade and credits. The relation STUDENT has records which have student id#, student name and date of admission.

There are also semantic properties attached to a relational database called *integrity constraints.* The most common type of integrity constraint is the *dependency constraint.* It is defined as a collection of fields in a record that uniquely determines other fields in a record. For example, in the TRANSCRIPT relation, STUDID determines STUDNAME and DEPT and COURSE# determine TITLE and CREDITS. In STUDENT, ID# determines NAME and DATEADM.

When a set of fields determines the rest of the fields in the row, that set is called a *key.* For example, ID# is a key for STUDENT. Also, the reader can check that STUDID, SEMESTER, YEAR, DEPT and COURSE# is a key for TRANSCRIPT.

Hierarchical Databases

In a *hierarchical database* data appears to the user as "hierarchies" of records. For example, Figure 2 contains an instance of a hierarchical database. The database represents registration information at a university. Each REGISTRAR record has two fields, SEMESTER and YEAR. Below each instance of a REGISTRAR record are instances of COURSE records which contain DEPT and COURSE# values. Below each COURSE record are SECTION records which contain SEC#, INSTRUCT, LOCAT and TIMES corresponding to the section. Finally, below each SECTION record are STUDENT records which contain STUDID, STUDNAME and GRADE values. In addition, there are STUDINFO records that contain ID#, STUDNAME, BIRTHDATE and DATEADM values.

In a hierarchy, a specific record can have sets of records below it called "children". Each of the children can themselves have sets of records below them as their children. A special characteristic of a hierarchical database is that once a record name is used at some level of the hierarchy, the same record name cannot be used anywhere else in the hierarchy.

TRANSCRIPT									
STUDID	STUDNAME	SEMESTER	YEAR	DEPT	COURSE #	SEC #	TITLE	GRADE	CREDITS
123456789	SMITH, ROBERT J.	FALL	1979	COMPSCI	103	001	INTRO	B	3
123456789	SMITH, ROBERT J.	FALL	1979	MATH	103	001	CALC	B	3
123456789	SMITH, ROBERT J.	FALL	1979	ART	101	001	INTRO	B	3
123456789	SMITH, ROBERT J.	FALL	1979	HIST	102	001	AM HIST	C	3
123456789	SMITH, ROBERT J.	SPRING	1980	COMPSCI	220	001	DISCR STRUCT	B	3
123456789	SMITH, ROBERT J.	SPRING	1980	MATH	104	001	CALC	B	4
123456789	SMITH, ROBERT J.	SPRING	1980	ECON	101	001	INTRO	C	3
123456789	SMITH, ROBERT J.	SPRING	1980	ENG	105	001	COMPOS	A	3
381258823	JONES, MARY T.	SPRING	1980	MATH	104	001	CALC	A	4
381258823	JONES, MARY T.	SPRING	1980	ECON	101	001	INTRO	C	3
381258823	JONES, MARY T.	SPRING	1980	PHYSICS	102	001	INTRO	D	4
381258823	JONES, MARY T.	SPRING	1980	ENG	105	001	COMPOS	I	3

STUDENT		
ID #	NAME	DATEADM
123456789	SMITH, ROBERT J.	SEPT 2, 1979
381258823	JONES, MARY T.	FEB. 1, 1980

Figure 1. A relational database.

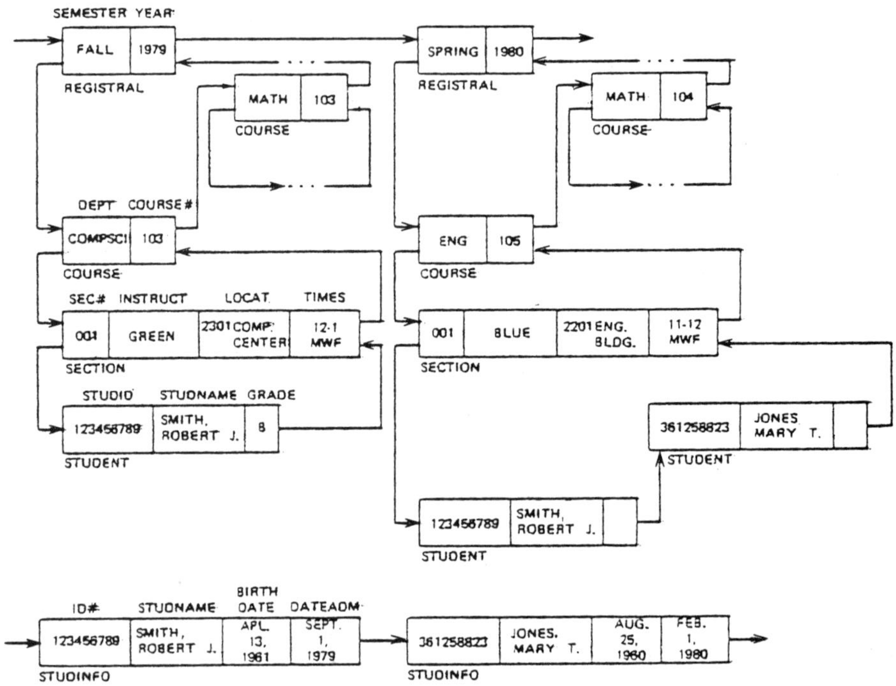

Figure 2. A hierarchical database.

Hierarchical databases also have integrity constraints. For example, in the REGISTRAR record, {SEMESTER, YEAR} is a key. That means that no two different REGISTRAR records in the entire database can have the same SEMESTER, YEAR pair. In the COURSE record, DEPT, COURSE# is a key. This means that below a particular REGISTRAR record (i.e., identified by a particular SEMESTER, YEAR) no two COURSE records can have the same DEPT, COURSE# pair. Similarly, in the SECTION, STUDENT and STUDINFO records, SEC#, STUDID and ID# are, respectively, also keys.

Network Databases

In a *network database,* data appears to the user as sets of records where each record instance may have many children and itself be the child of many parents. Whereas, in the hierarchical approach, a child record can have only one parent, in the network approach, a child record can have many parents. In the hierarchical approach, a record type can appear only once in its ancestrial tree. In the network, this restriction disappears.

For example, Figure 3 contains an instance of a network database. The database represents catalogue information for courses at a university. The record type CATALOGUE contains fields DEPT, COURSE#, TITLE, CREDITS, and DESCR (description). Each CATALOGUE record has below it (called a CATALOGUE-LINK "set") a single occurrence of a LINK record with no data. Below each LINK record (in a LINK-CATALOGUE set) are other occurrences of the CATALOGUE record. The lower level CATALOGUE records indicate the courses that are prerequisites for the courses in the CATALOGUE records at the upper level.

Although network systems allow the same record type to occur several places in the anscestral structure (e.g., CATALOGUE), many do not allow both parent and child to be of the same type. This is the reason for the existence of the intermediary LINK record in our example. As in the relational and hierarchical cases, network databases can also have integrity constraints on the data. For example, in CATALOGUE, fields DEPT and COURSE# form a key. Since LINK has no data, it has no key.

2. MULTI-DATABASE ISSUES

In this section, we will discuss several important database issues. The reader will observe that there is a fundamental theme linking all of these issues; namely, they all involve two or more different databases.

External-to-Conceptual Mapping

Suppose that a university maintains the hierarchical database of Figure 2. Suppose also that a group of students prefer to see the data in terms of the relational database of Figure 1. A solution is to build the relational database as an outside or "external" view of the underlying "conceptual" hierarchical database. That is, the students can pretend that the database is relational even though, in fact, it is hierarchical.

The *external-to-conceptual mapping problem* is as follows. How can one precisely express the relationship between two databases so that one can be maintained as an "external" view of a second "conceptual" database? The mapping should provide a mechanism by which programs at the external level can be transformed into programs at the conceptual level. It should also provide a means by which data can be transferred back and forth between levels.

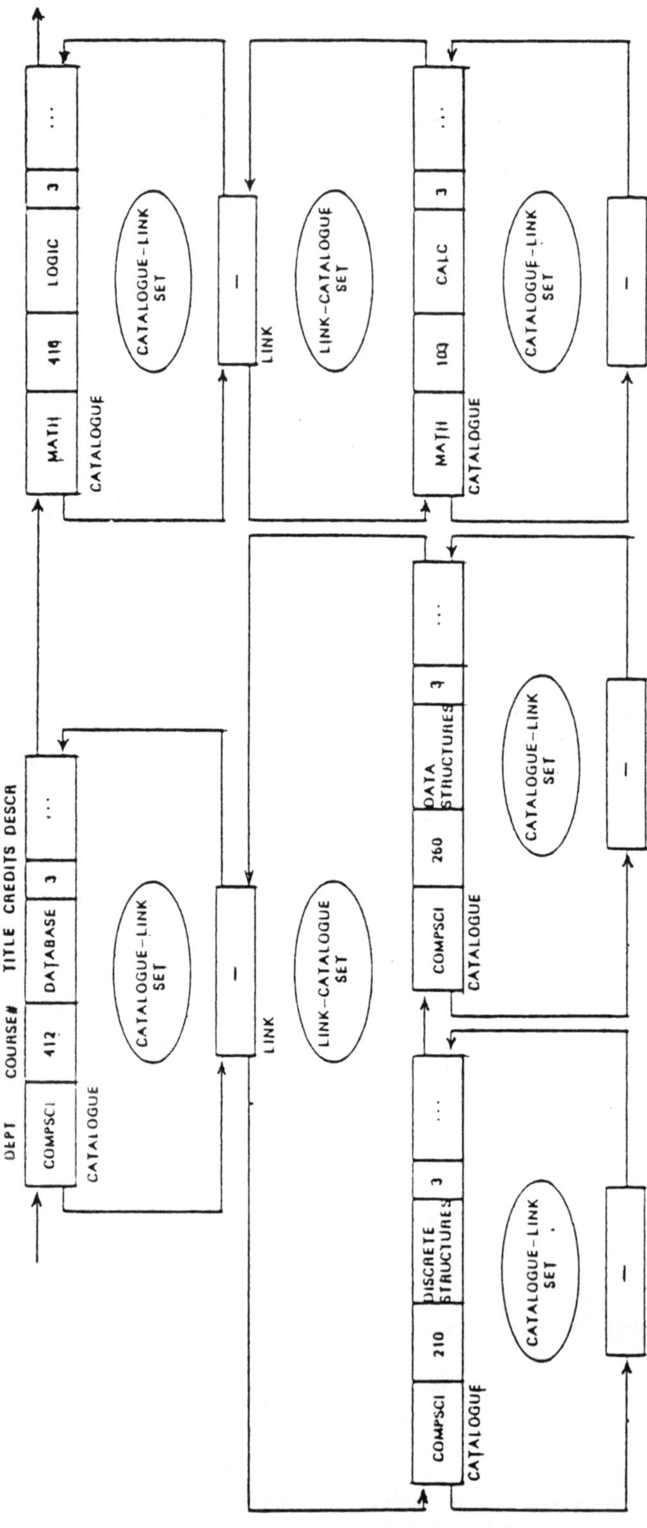

Figure 3. A network database.

View Update

Suppose that, in some way, the student's relational database was maintained as an external view of the hierarchical database. Then, as far as the students are concerned, there is only one database; namely, the relational one. Suppose that a student wants to insert a new row into the TRANSCRIPT table. Since the TRANSCRIPT table does not physically exist, this insertion has to be "simulated" on the underlying hierarchical database.

The *view update problem* is that of determining how updates (e.g., insertions, deletions, etc.) on an external view can be simulated "correctly" on the underlying conceptual database. Constraints also play a role in the view update problem. For example, suppose an update is desired at the external level which would violate an external constraint (e.g., a key constraint). Then the update simulation at the conceptual level should, in some way, catch the violation and prevent it from occurring.

View Integration

Suppose that a university database is being planned or designed. Suppose also that in the analysis, it is discovered that there are three classes of users- the registrar who wants a hierarchical database like that of Figure 2, the students who want a relational database like that of Figure 1 and the departments which want a network database like that of Figure 3. The problem is how to build a database that will satisfy the needs of all these potential users.

The *view integration problem* is as follows. Given a collection of views, how can one construct a conceptual database that will support all of them as external views? The choice of underlying database should be such that the most frequent users incur the least overhead in the translation between external and conceptual levels. The view integration problem is often a major component in a design methodology in which external views are first extracted from a community of potential users and then integrated into a conceptual database.

Database Conversion and Query Processing

Suppose that the university designs its database so that the underlying conceptual database is the hierarchy of Figure 2 plus the network of Figure 3. Thus, the registrar, the students and the departments all have their own views established. Suppose that the students no longer want to pay the price of the overhead of translation into the conceptual hierarchy. A solution is to build another copy of the database in the form of the relational view (Figure 1) that they use. Thus, one is giving up centrality of the data for improved perfomance for a class of users.

The *database conversion problem* is that of finding a methodology by which already existing databases can be used to construct new databases. The result of processing a query is often a "table" of values that satisfy the query. The problem of *query processing* is actually a special case of the database conversion problem. This is because in query processing, we are constructing new "relational" databases from other databases.

Automatic Program Conversion

Suppose that in time, the students decide to change their relational database to the hierarchical database of Figure 2. During the period that they used their relational database, they collected a large (and costly) amount of software. Unfortunately, the old software refers to the old database (i.e., TRANSCRIPT and STUDENT tables), hence, cannot be used on the new hierarchical database. Consequently, all the old software has to be discarded and new software has to be written.

The *automatic program conversion problem* is that of deriving, from the old and new database descriptions, a mechanical means for transforming old application programs to programs on the new database. The process not only involves translation from the old to the new, but also optimization of the resultant programs. Furthermore, in some cases, the language in which the old programs are written is different from the language of the new database. This would, indeed, be the case if the old and new databases are in different database management systems.

External Axiomatization

Suppose that the student's relational database of Figure 1 is maintained as an external view of the hierarchical database of Figure 2. Thus, every instance of the conceptual database "produces" an instance of the external relational database. Furthermore, the constraints on the data in the conceptual database (e.g., DEPT, COURSE# determines TITLE and CREDITS in CATALOGUE) reflect the properties of the data in the produced external database (e.g., DEPT, COURSE# determine TITLE and CREDITS in TRANSCRIPT).

The *external axiomatization problem* is that of finding a characterization of all possible external databases. Such a characterization is important, especially if updating on the external view is permitted. Consequently, an external user can know before hand if the results of an external update are within the range of possible external databases.

Hidden Databases

The notion of one database "hidden" in a second database is an important commonality between all the issues mentioned in the previous section. For example, in the external-to-conceptual mapping problem, the question is how to "hide" the external database within the conceptual database. In view update, the problem is how to simulate updates on the hidden external database by "correct" operations on the conceptual database. In view integration, the problem is how to choose a single database that can hide a set of different databases.

In database conversion, the problem is how to construct a database that is hidden within the old database. In automatic program conversion, the old database is thought of as being hidden in the new database. The problem then is how to convert programs to the new database so they will have the same effect as the old programs have on the hidden database. In external axiomatization, the problem is to characterize the class of all databases that are hidden within instances of the conceptual databases.

3. LOGIC AND DATABASES

One of the benefits of working with the relational approach to databases is that it can be couched within the formalism of first-order predicate logic. As a result a mathematical foundation is available for dealing with database issues when databases are all relational.

Relational Databases in First-Order Logic

By a *first-order theory,* we mean a set of formulae or statements in some formal language. For example, recall the constraint on the database of Figure 1 which says that in the STUDENT table, ID# determines NAME and DATEADM. This is expressed in first-order logic by the following statement:

(STUDENT(ID#:1,NAME:1,DATEADM:1)
& STUDENT(ID#:1,NAME:2,DATEADM:2))
->(NAME:1=NAME:2 & DATEADM:1=DATEADM:2).

That is, if two rows appear in the STUDENT table with the same ID#, then they will also have the same values for NAME and DATEADM.

As another example, recall the constraint which says that in the TRANSCRIPT table, STUDID determines STUDNAME. This is expressed in first-order logic as:

((TRANSCRIPT(STUDID:1,STUDNAME:1,SEMESTER:1,...,CREDITS:1)
& TRANSCRIPT(STUDID:1,STUDNAME:2,SEMESTER:2,...,CREDITS:2))
->(STUDNAME:1=STUDNAME:2).

A theory that represents the database of Figure 1 would consist of all the statements, like the above, which represent constraints on the data.

By a *first-order structure,* we mean a mathematical entity which has 1) a universe of objects, 2) a set of predicates (i.e., each of which evaluates to true or false) and 3) a set of functions on the universe. For example, a first-order structure corresponding to the relational database in Figure 1 is as follows. A universe would be comprised of objects such as EBCDIC DECIMAL (9) values for STUDID plus EBCDIC CHAR(20) values for STUDNAME,

Two predicates would correspond to the TRANSCRIPT and STUDENT tables. For example,

STUDENT(123456789,SMITH ROBERT J.,SEPT 2 1979)

is "true" because this row is in the STUDENT table in Figure 1. A function TOTCREDITS would be one which inputs a studid, semester and year and returns the total number of credits taken by that student during that semester year, e.g., according to Figure 1,

TOTALCREDITS(123456789,FALL,1979)=12.

By a *model* of a first-order theory we mean a first-order structure which satisfies all the statements in the theory. **At any moment of time, a relational database can be regarded as a model of a first-order theory, namely, its constraints.** For example, the instance of the relational database pictured in Figure 1 does not violate any of the constraints in the theory described above. Hence, the instance is a model of that theory.

In the previous section, several issues were discussed which involved two or more databases. From the above we see that a relational database can be represented as a theory in first-order logic. Consequently, these issues reduce to, in the relational case, issues between two or more theories. To get a grip on such problems, let us see how multiple theories are handled in first-order logic.

Interpretations in First-Order Logic

In ordinary first-order logic, whenever two or more theories are contrasted, the usual comparison operator is the "interpretation". This notion will play a fundamental role throughout this book. The concept of an interpretation can be illustrated by considering an interpretation I(R;C) of the theory of real numbers R in the theory of the complex numbers C.

First, we associate with every real number in R a complex number, called its "code" in C. For example, we code the real number 3.5 by the complex number 3.5+0i . Second, we associate with predicates and functions in R representative predicates and functions in C that work on the codes. For instance, we represent real equality "a=b" by the complex equality on codes "(a+0i)=(b+0i)". Also, we associate with real addition "a+b" the complex addition on codes "(a+0i)+(b+0i)=(a+b)+0i". Third, each of the axioms of the theory of the real numbers is "implied" from the axioms of the theory of the complex numbers. For example, the commutitivity of complex addition (e.g., (a+bi)+(c+di) = (c+di)+(a+bi)) implies the commutitivity of addition on the codes (e.g., (a+0i)+(b+0i) = (b+0i)+(a+0i)).

When one theory is interpreted in a second, then any model of the second theory has "hidden" inside of it a model of the first. For example, in the interpretation I(R;C) of R in C given above, the model of the real numbers is hidden inside of the model of the complex numbers. First, the set of real numbers is hidden inside the set of complex numbers. Namely, each real number "a" is hidden as the complex number "a+0i". Second, the real predicates and functions are hidden inside of complex predicates and functions. For example, the real equality predicate "a=b" is represented by the complex equality predicate as "(a+0i) = (b+0i)". Also, the real addition function "a+b" is hidden inside of the complex addition function as "(a+0i)+(b+0i) = (a+b)+0i". Third, since each of the axioms of R is "implied" by the axioms of C, the hidden reals will also be a model of the axioms in R.

Database Logic Views and Interpretations

In the book APPLIED DATABASE LOGIC I: FUNDAMENTAL DATABASE ISSUES, the formal system called *database logic* is introduced. Database logic is to the relational, hierarchical and network approaches as first-order logic is to the relational approach. That is, it provides a mathematical foundation for the study of databases. The book is relatively self-contained. Consequently, little previous knowledge of logic (first-order or database) is required for understanding the presentations.

The notion of a first-order theory generalizes to the notion of a *database view* in database logic. The notion of a first-order structure generalizes to the notion of a *database structure*. **At any instant of time, a database (relational, hierarchical or network) is regarded as a database structure that is a model of the view.** Since the notion of database view subsumes that of first-order theory, all results obtained using database views give us, as a special cases, results about relational databases.

The notion of an interpretation in first-order logic generalizes to the notion of *interpretation of database views* in database logic. Roughly speaking, an interpretation of one database view in a second is a correspondence between the objects (i.e., data, functions, predicates, etc.) of the first view and those of the second. In addition, the constraints in the first view are "implied" from the constraints in the second view. Since the notion of database view interpretation subsumes that of first-order interpretation, all results obtained using view interpretations give us, as a special cases, results about relational databases.

4. RELATED READING

Date, C. J., *An Introduction to Database Systems, Volume I (third edition)* Addison-Wesley Publishing Company, Reading, Massachusetts, 1983.

Kroenke, D., *Database Processing*. Science Research Associates, Inc., Chicago, Illinois, 1977.

Martin, James, *Computer Data-Base Organizations*. Prentice-Hall, Englewood Cliffs, New Jersey, 1975.

Sibley, E., (editor) *Data-Base Management Systems.* ACM Computing Surveys, vol 8, March 1976.

Ullman, J.D., *Principles of Database Systems (second edition).* Computer Science Press, Rockville, Maryland, 1982.

APPLIED DATABASE LOGIC I:

FUNDAMENTAL DATABASE ISSUES

PRENTICE-HALL, INC., ENGLEWOOD CLIFFS, NEW JERSEY

CONTENTS

Proceedings of Symposia in Applied Mathematics
Volume 34, 1986

Data Compression Algorithms

Victor S. Miller

ABSTRACT Compression of data is important both for lowering the cost of communication, and for lowering the cost of storage. There have been innumerable techniques for compression dating back at least as far as Morse Code. There are two main classes of techniques: One views the message to be compressed as a typical member of a class of messages. The other tries to work on the individual message. A survey of the ideas and significant results will be given with an emphasis on algorithms.

1. Introduction. The compression of messages is quite an old technique. With the advent of relatively inexpensive processing power, the use of compression should become quite common. This paper will concentrate on algorithms and techniques for what is called *Noiseless and Lossless compression.* That is, the medium recording the data is assumed not to introduce errors, and that it is possible to decode the compressed data to obtain the original data back again. The literature on compression is quite voluminous. The bibliography of this paper includes what, in the author's opinion, is a representative sample of the major results.

1.1. Terminology and Notation. In this paper we shall be talking about strings of symbols. A *string* is a sequence of *symbols* of a given alphabet (finite set). If σ and τ are strings then $\sigma\tau$ denotes the string formed by following the sequence of symbols in σ with the sequence in τ. This is also called the *concatenation* of the two strings. If σ is a string then $|\sigma|$ denotes the length of the sequence (string). The *empty string* is the string of length 0, and is denoted by Λ . We denote $\log_2 x$ by $\lg x$.

1.2. The specific problem. The abstract problem is the following:
We assume that there is an *encoder* and a *decoder*. The encoder transmits a string of symbols to the decoder, who performs some processing and produces a decoded string. Given an alphabet, Σ, (i.e. a finite set) of symbols, we are given

1980 Mathematics Subject Classification. 94A29.

Key Words and Phrases. Compression, Entropy, Algorithms.

messages which are strings of symbols from the alphabet. The problem is to encode the string into a string over another alphabet , B , usually taken to be the set {0,1} of binary digits, in such a way as to minimize (or reduce) the cost of the encoded string. The cost of a string is the sum of the costs of its individual symbols, which are assumed known (and greater than 0). The only practical encoding methods are those which are algorithms: Those procedures which are guaranteed to halt on every input.

We can be more specific in the above description: Both the encoder and decoder perform as follows: Each reads a symbol from input, does some computation, and emits a (possibly empty) string of symbols. We further assume that the alphabet is augmented by a distinguished symbol "#", which will indicate end of input. The encoder and decoder may be thought of as defining functions on strings: $E(\sigma)$ and $D(\sigma)$, satisfying $E(\sigma\alpha) = E(\sigma)\beta$ where β is the string emitted by the encoder, having already read σ , after reading the symbol α. The analogous statement holds for the decoder. We assume that no other information is transmitted from the encoder to the decoder.

We further say that a coding algorithm has *delay d* if the coder does not need to look more than d symbols ahead in the input before emitting a piece of the code sufficient to decode the current symbol. In terms of the above functions this means that if σ is a string, α a symbol, and δ is any string of length d, then $\sigma\alpha$ is a prefix of $D(E(\sigma\alpha\delta))$. We also say that an algorithm is *off-line* if it needs to look at the entire input before emitting any output. For most purposes, practical coding algorithms are those with delay d, for small d. We say that the *information* in a string σ relative to a given encoding E is $| E(\sigma\#) | / | \sigma\# |$ the average number of output bits per input symbol.

1.3. Entropy and Information. The algorithms used fall into two main classes: Coding of *individual sequences*, which means that only the input is considered in trying to get the best encoding. This is opposed to *source coding*, which tries to find an encoding which is best (on the average) for a certain set of strings, where each string is assigned some probability. In *universal coding* a certain parametrized class of probability distributions is postulated and that member of the class which fits the input best is chosen. A source coding method approriate to that member is then used. A (Markov) *source* is something which emits symbols according to some probability distribution. The most common are sources with finite memory: The probability distribution depends upon the previous n symbols emitted, for some fixed n. The *entropy* of a source is defined to be

$$H(p) = -\sum_i p_i \lg p_i$$

where p_i is the probability of the i-th event, usually taken to be the emitting of the i-th symbol in the alphabet. If we are dealing with a n-th order source the probabilities are conditional upon the previous n symbols. The entropy of the source is then the expected value of the conditional entropy.
Entropy is important because it gives a lower bound on the average information in a string emitted by a source relative to any encoding algorithm. See Khinchin ([31]) for details.

2. Codes. There are three main types of codes used for compression: *concatenation codes*, *block codes*, and *arithmetic codes*. In the first kind, individual symbols (or "extended symbols", see below) are encoded one at a time, by looking up the symbol in a *code book*, (i.e. a table of codes corresponding to symbols) which may depend on the symbol's context, and emitting that code. In the second kind, the input message is divided into blocks of symbols which are then encoded by some means. The code for the message is the concatenation of the codes for all of the blocks. There is also a third type of code, the so-called "arithmetic code" which encodes the whole message as a unit, but has the property that only a small fixed amount of the code needs to be worked on at a time. An example of the first kind is the *Huffman code* ([26]). An example of the second kind is the *Shannon-Fano code* ([67] , [17] , [18]). An example of the third kind is the *Elias code* ([19]).

2.1. Prefix Codes. We need some terminology when speaking about concatenation codes. We assume that we have a source alphabet $\{X_1, \ldots, X_n\}$. A *code* is a set of code words, $\{C_1, \ldots, C_n\}$, which are strings over a code alphabet. The source message $X_{i_1} \ldots X_{i_k}$ is encoded as $C_{i_1} \ldots C_{i_k}$. An *encoded message* is the concatenation of one or more code words. A code is said to be *uniquely decipherable* or UD, if each encoded message is an encoding of a unique source message. A *prefix code* is a code in which no code word is the prefix of another. A prefix code is trivially UD. A code is *exhaustive* if any string is either a prefix of a code word, or has a code word as a prefix.

2.2. The Huffman Code. The first successful practical code, which is still in wide use, is the *Huffman Code*. This code solves the following optimization problem:

> Given a set of probabilities $p_1, \ldots, p_n$ whose sum is 1, find code words $C_1, \ldots, C_n$ such that $p_1 \mid C_1 \mid + \cdots + p_n \mid C_n \mid$ is a minumum.

The code $\{C_1, \ldots, C_n\}$ is required to be UD. Huffman solved this optimization problem, in the case of binary output alphabet, by means of a transforming the problem into one about trees and using a "greedy algorithm":

1. Create nodes for each input symbol, weighted by its probability.
2. Find the two parentless nodes of lowest weight and form a new parent node whose weight is the sum of the two weights of its children.

3. Continue step (2) until only one node is left without a parent node

4. Each node corresponding to an original symbol has a unique path from the root of the resulting tree. The code for that symbol is the string of 0's and 1's which gives the path from the root: 0 meaning go left, and 1 meaning go right.

The search in step (2) may be made efficiently by using the structure of a *priority queue*. See [1] , pp. 147-152, for details.

2.3. The Shannon-Fano Code. The Shannon-Fano code was one of the first codes proposed. It works in the following way: The input message is divided into fixed length blocks of length l. All of the possible messages of length l are ordered in order of their increasing probability. Say the messages are $M_1, \dots, M_n$ with corresponding probabilities $p_1, \dots, p_n$. Let P_j denote the cumulative probability $p_1 + \cdots + p_{j-1}$. Then the encoding of message j is the first k bits in the binary expansion of P_j where $k = -[\lg p_j]$. Although simple to describe, there are practical objections to this code. In order to do well, the block length, l, needs to be at least of moderate size. This entails the consideration of a large number of messages.

2.4. The Elias Code. One of the problems with the first two mentioned codes is that the information relative to the encoding is usually larger than the entropy of the source. In order to approach the entropy, the block length needs to increase for the Shannon-Fano code, and blocks of symbols, instead of individual symbols, need to be encoded with the Huffman code. This is undesirable from the point of view of computation because of the explosion in the amount of storage and processing necessary. This problem was solved in theory by the Elias Code. Later workers modified the code slightly so that it was amenable to finite computation and even efficient. See the section on Arithmetic Codes for details. It is possible to view code words in a concatenation code like the Huffman code as numbers: If $C = a_1 a_2 \dots a_n$ is a code word, then define $F(C) = a_1 s^{-1} + a_2 s^{-2} + \cdots + a_n s^{-n}$. In the above expression s is the number of symbols in the input alphabet, and each symbol a_i stands for a distinct integer between 0 and $s - 1$. Then we have the following properties:

1. If $\{C_1, \dots, C_n\}$ is a prefix code, then $F(C_1), \dots, F(C_n)$ are distinct.

2. If the code in question is an exhaustive prefix code, and $F(C_1) < \cdots < F(C_n)$ then $F(C_i) = s^{-l_1} + \cdots + s^{-l_i}$ where $l_i = |C_i|$.

3. The Kraft-McMillan-Szilard inequality

$$s^{-l_1} + \cdots + s^{-l_n} \leq 1$$

4. If the code is an exhaustive code then above inequality is an equality.

5. The Huffman code may be thought of as estimating the probabilty p_i of the i-th symbol by s^{-l_i}.

Because of (4), the probabilites add up to 1. The $F(C_i)$ are just the cumulative probabilities of the symbols. The average symbol length under the Huffman code is exactly equal to the entropy with the above set of estimated probabilities. In numerical terms, a Huffman coder performs steps such as the following:

$$S \leftarrow S + PF(C_i)$$
$$P \leftarrow Pp_i$$

where S is the encoded output. The variables S and P are initialized to 0 and 1 respectively. The Elias code does exactly this, relaxing the restriction that p_i be a power of 2.

In the Elias code we assume that we can determine the probabilities of emitting any symbol at any point. Choose an arbitrary ordering of the symbols in the alphabet (there in fact may be one already given). For each assignment of probabilities to symbols, associate with each symbol α_i , whose probability is p_i, a half-open interval $[P_i, a_i + p_i)$ in $[0, 1)$, where $P_i = p_1 + \cdots + p_{i-1}$. Now, inductively, we associate with each string σ an interval $[a_\sigma, a_\sigma + L_\sigma)$ as follows:

$$a_\Lambda = 0$$
$$L_\Lambda = 1$$
$$a_{\sigma\alpha_i} = a_\sigma + P_i L_\sigma$$
$$L_{\sigma\alpha_i} = p_i L_\sigma$$

This assignment has the property that if two strings are not prefixes of one another, then their associated intervals are disjoint. The code for the string σ is the first k bits in the binary expansion of the number $a_\sigma + L_\sigma/2$ where $k = -\lfloor \lg L_\sigma \rfloor$.

2.5. *Arithmetic Coding.* A number of authors ([51] , [28] , [65] , [9] , [25] , [59] , [56] , [58] , [38]) have done work which makes the Elias code practical. They basically use the same algorithm as Elias along with the techniques of scaling and rounding used in floating point arithmetic so that only a fixed length part of the numbers involved is used in arithmetic operations. Langdon and Rissanen have called such codes *Arithmetic Codes.* (Note: the term Arithmetic Code is also used with another meaning in the coding literature. In that context it means an error-correcting code used for arithmetic operations)

2.6. *Modeling and Universal Coding.* One of the great practical problems with source coding is in determining appropriate characteristics of the source. An interesting discussion of this problem from the standpoint of encoding the printed Enlish Language is contained in papers of Shannon ([68]) and Cover and King ([10]). A very common practice is to fix a certain class of probabilistic sources and to use some characteristics of the input string to choose that member of the class which somehow best represents the given input string. *Universal Coding* was invented by Kolmogorov to address this problem. To quote Elias ([14]) "A universal code is a code that works for all sources in some class" To be more specific, we have Davisson's definition ([11])

> Universal coding is any method of block encoding for composite sources where the following holds: a) The encoding depends only on the observed source message encoded and not past or future outputs, i.e. it is blockwise memoryless. b) Some performance measure(s) is attained arbitrarily closely as the block length goes to infinity

Davisson specifically excludes *adaptive coding* in which the estimates of the sources probabilities are constantly revised (adapted) depending upon the string observed so far. Davisson does this for techncal reasons, but other authors seem to adhere more to Elias' view. In Kolmogorv's formulation the source is divided into fixed-length blocks. Within each block some computation is done to determine the best match to the current block from the considered class of sources. The encoding for the block then consists of an encoding of the choice among the sources considered, followed by the encoding of the block according to an optimal code for that source. As the class of sources considered gets larger and larger, more and more of the block is taken up with the encoding of the choice among the sources, and less and less is taken up with the actual encoding of the input. It was probably this phenomenon which led Kolmogorov to formulating his definition of *algorithmic entropy*. To give a specific example, suppose that the input string consisted of an indefinite number of repetitions of the string 01, and that the class of sources consisted of all n-th order Markov sources. Since most probabilities involved would be 0 when n was large, we might use some sort of sparse encoding, which would essentially be a description of those substrings which were present, possibly in terms of shorted substrings. Once we had done this, there would be very little else to specify.
The problem of good modeling is still more of an art than a science with respect to chosing an approriate class of sources to consider. The papers of Rissanen and Langdon ([37] , [60]) provide a good discussion of the problems involved.

3. Algorithmic Information. Kolmogorov and Chaitin independently developed information theory based on the concept of *algorithmic complexity*. Given a fixed sufficiently powerful model of computation (a universal Turing Machine) one says that the information in a string is the length of the smallest input to the given machine which will produce the given string. For further details see [7] . Even though theoretically satisfying, a problem with the definition is that the information of a string is not computable. For attempts at solutions to this problem, see the next section.

3.1. The Coding Methods of Ziv and Lempel. Ziv and Lempel, in a series of papers ([41] , [75] , [76] , [77]) describe two methods of coding individual sequences which are based on algorithmic information theory. Unlike the general measures of algorithmic information described in the last section, which are not computable, Ziv and Lempel minimize over a class of automata which is small enough to allow the minimum to be computed, at least in principle. They further prove that their methods are *universal* in the sense that if their encoding algorithm is used on a class of finite order Markov sources, then the bits per symbol used in the output converges to the standard entropy, for almost all members of the class. See [77] for details. The first Ziv-Lempel code works as follows:

1. Starting at the current position in the input, find the longest string which occurs somewhere earlier in the input (at the beginning, this string is the empty string).
2. Output an encoding of (position,length) of the matched string followed by an encoding of the first input symbol after the matched string.
3. Advance the input pointer past the matched string and additional symbol, and return to step (1) if not at the end-of-string.

This encoding is clearly computable in time $O(N^2)$ where N is the length of the input string. Ziv and Lempel prove that the information relative to this encoding has many of the general desirable properties for an information measure. For details see [76] . The work of Rodeh, Even and Pratt ([61]) showed that the above encoding could be accomplished in time linear in the size of input string (i.e. a constant time per input symbol).

Ziv and Lempel also gave a second method of encoding which is simpler than the first, which also has the same general desirable properties as an information measure. In fact they show that if one restricts the class of encoders to those whose computation is done by a finite state automaton (i.e. the encoder has only a fixed amount of memory) one can define an algorithmic entropy measure by minimizing over a fixed size memory and taking limits. This measure then has all of the usual properties of an information measure, and coincides with ordinary entropy when given the output of a source to encode, in the sense that if one takes the expected value of the measure over all strings produced by the source then the entropy of the source is attained. The restriction to computation by a finite automaton is justified in practice because virtually all encodings used fit into that class.

1. Initialize a set of strings (the "dictionary") to contain the empty string. Set N=1
2. Starting at the current position in the input, find the longest string which is a member of the dictionary.
3. Output an encoding of the identifying number for the matched string followed by an encoding of the first input symbol after the matched string.

4. Add a new string to the dictionary consisting of the matched string plus the additional symbol. Set N=N+1. Give the new string identifying number N.

5. Advance the input pointer past the matched string and additional symbol, and return to step (2) if not at the end-of-string.

The encoding for the identifying number mentioned in step (2) would be k bits of the binary representation of the number, where $k = [\lg N] + 1$. This may be implemented in time linear in the length of the input and space proportionaly to the encoded string by using a TRIE structure (see Knuth [32]).

3.2. The Encoding Algorithms of Miller and Wegman. Miller and Wegman ([46] , [3]) have improved the second encoding algorithm of Ziv and Lempel in three respects: a) No encoding of the additional symbols in the input is required b) The Ziv-Lempel algorithm has no fixed bound on the requirement for storage. A way to make use of a small amount of storage efficiently is given. c) A new method is given for adaptation (step 3) which usually results in faster adaptation. The first modification follows:

1. Initialize a set of strings (the "dictionary") to contain the all single symbols. Set N=the size of the input alphabet.

2. Starting at the current position in the input, find the longest string which is a member of the dictionary.

3. Output an encoding of the identifying number for the matched string.

4. Add a new string to the dictionary consisting of the matched string plus the additional symbol. Set N=N+1. Give the new string identifying number N.

5. Advance the input pointer past the matched string (but not the additional symbol), and return to step (2) if not at end-of-string.

The encoding of the identifying number is the same as in the second Ziv-Lempel algorithm.

The method for making use of a small fixed amount of storage consists in deleting strings from the dictionary, and re-using their identifying numbers. The principle of *least recently used* is used. This basically uses the philosophy of adaptation to predict which strings are not likely to be used on the basis of past input. Sleator and Tarjan ([69]) have shown that such a strategy is only a constant factor from optimal.

The third improvement is in the strategy for adaptation. In the original algorithm a new string added to the dictionary has length only one greater than a the matched string. Thus new strings may grow rather slowly. In the third modification, a new string is allowed to be the concatenation of two strings in the dictionary. This modification may also be combined with the replacement strategy cited above. The third improvement follows:

1. Initialize a set of strings (the "dictionary") to contain the all single symbols. Set N=the size of the input alphabet. Set U=the empty string.

2. Starting at the current position in the input, find the longest string which is a member of the dictionary. Set S=this string.

3. Output an encoding of the identifying number for the matched string.

4. Add the string US to the dictionary. Set N=N+1. Give the new string identifying number N. Set U=S.

5. Advance the input pointer past the matched string (but not the additional symbol), and return to step (2) if not at end-of-string.

4. Summary. The topic of data compression is much greater than could be touched on in this brief survey. The bibliography provides a number of references in other areas. Some of the most challenging areas are in coding of two dimensional data (such as television images) and in coding for constrained channels (not every string in the coding alphabet is permitted).

BIBLIOGRAPHY

1. A. V. Aho, J. E. Hopcroft and J. D. Ullman, *The Design and Analysis of Computer Algorithms*, Reading, Addison-Wesley, 1974.

2. R. Arnold et. al., *Markov Processor for Context Encoding from Given Characters and for Character Decoding from Given Contexts*, U. S. Patent 4,099,257, July 4, 1978.

3. M. A. Auslander, W. H. Harrison, V. S. Miller and M. N. Wegman *PCTERM: A terminal emulator using compression*, to appear, Proceedings of GLOBECOM '85, IEEE press.

4. H. Blasbalg and R. Van Blerkom, *Message Compression*, IRE Trans. on Space Electronics and Telemetry, *SET-8* (62), 228-238

5. G. J. Chaitin *On the Length of Programs for Computing Finite Binary Sequences*, JACM, *13* (66), 547-569

6. G. J. Chaitin *A Theory of Program Size Formally Identical to Information Theory*, JACM, *22* (75), 329-340

7. G. J. Chaitin *Algorithmic Information Theory*, IBM J. Res. Dev., *21* (77), 350-359

8. D. Cooper and M. F. Lynch, *Text Compression Using Variable- to Fixed-Length Encodings*, J. Amer. Soc. for Info. Science, *33* (82), 18-31

9. T. M. Cover, *Enumerative Source Coding*, IEEE Trans. Inform. Theory, *IT-19* (73), 73

10. T. M. Cover and R. C. King, *A Convergent Gambling Estimate of the Entropy of English*, IEEE Trans. Inform. Theory, *IT-24* (78), 413-421

11. L. D. Davisson, *Universal Noiseless Coding*, IEEE Trans. Inform. Theory, *IT-19* (73), 783-795

12. Y. Desmedt *Parallel coding of a discrete source*, Preprint

13. Y. Desmedt, J. Vanderwalle and R. Govaerts *A high speed parallel discrete source coder*, Picture coding Symposium 1981, June 3-5, 1981, Montreal, Canada, pp. 58-59.

14. P. Elias, *Universal Codeword sets and representation of the integers*, IEEE Trans. Inform. Theory, *IT-21* (75), 194-203

15. P. Elias, *Minimax Optimal Universal Codeword Sets*, IEEE Trans. Inform. Theory, *IT-29* (83), 491-502

16. S. Even, *On Information Lossless Automata of Finite Order*, IEEE Trans. Electronic Computers, *EC-1* (65), 561-569

17. R. M. Fano, *Technical Report no. 65*, Res. Lab. for Electronics, MIT, Cambridge, 1949.

18. R. M. Fano, *Transmission of Information*, New York, John Wiley, 1961.

19. R. G. Gallager, *Information Theory and Reliable Communications*, New York, John Wiley, 1968.

20. R. G. Gallager, *Variations on a Theme by Huffman*, IEEE Trans. Inform. Theory, *IT-24* (78), 668-674

21. E. N. Gilbert and C. L. Monma *Multigram Codes*, IEEE Trans. Inform. Theory, *IT-28* (82), 346-348

22. E. N. Gilbert and E. F. Moore, *Variable-length Binary Encodings*, Bell Sys. Tech. J., *38* (59), 933

23. S. W. Golomb, *Run-Length Encodings*, IEEE Trans. Inform. Theory, *IT-12* (66), 399-401

24. R. M. Gray and L. D. Davisson, *The Ergodic Decomposition of Stationary Discrete Random Processes*, IEEE Trans. Inform. Theory, *IT-20* (74), 625-636

25. M. Guazzo, *A General Minimum-Redundancy Source-Coding Algorithm*, IEEE Trans. Inform. Theory, *IT-26* (80), 15-25

26. D. A. Huffman *A method for the construction of minimum redundancy codes*, Proc. IRE *40* (52), 1098-1101

27. F. Jelinek and K. S. Schneider, *On Variable-Length-to-Block Coding*, IEEE Trans. Inform. Theory, *IT-28* (72), 765-774

28. C.B. Jones, *An Efficient Coding System for Long Source Sequences*, IEEE Trans. Inform. Theory, *IT-27* (81), 280-291

29. R. M. Karp, *Minimum-Reduncancy Coding for the Discrete Noiseless Channel*, IRE Trans. on Inform. Theory *IT-7* (61), 27-38

30. D. Kazakos, *Robust Noiseless Source Coding Through a Game Theoretic Approach*, IEEE Trans. Inform. Theory, *IT-29* (83), 576-583

31. A. I. Khinchin, *Mathematical Foundations of Information Theory*, Dover, New York, 1957.

32. D. E. Knuth, *The Art of Computer Programming, volume 3: Sorting and Searching*, Reading, Addison-Wesley, 1973.

33. A. N. Kolmogorov, *Three Approaches to the Quantitative definition of Information*, Problems of Information Transmission *1* (65), 1

34. A. N. Kolmogorov *Logical Basis for Information Theory and Probability Theory*, IEEE Trans. Inform. Theory, *IT-14* (68), 662

35. J. M. Kusmiss, *An Experiment in Adaptive Coding*, IBM Poughkeepsie, TR00.2514, March 21, 1974.

36. G. Langdon, *A Note on the Ziv-Lempel Model for the Compression of Individual Sequences*, IEEE Trans. Inform. Theory, *IT-29* (83), 284-287

37. G. G. Langdon, Jr. and J. Rissanen, *Compression of Black-White Images with Arithmetic Coding*, IEEE Trans. Comm., *COM-29* (81), 858-867

38. G. G. Langdon, Jr. and J. Rissanen, *A Simple General Binary Source Code*, IEEE Trans. Inform. Theory, *IT-28* (82), 800-803

39. G. G. Langdon, Jr. and J. Rissanen, *A Double Adaptive File Compression Algorithm*, IEEE Trans on Comm., *COM-31* (83), 1253-1255

40. E. Lawler, *Combinatorial Optimization: Networks and Matroids*, New York, Holt, Rinehart and Winston, 1976.

41. A. Lempel and J. Ziv, *On the Complexity of Finite Sequences*, IEEE Trans. Inform. Theory, *IT-22* (76), 75-81

42. L. S. Loh, J. H. Mommens and J. Raviv, *Method of Achieving Data Compaction Utilizing Variable-length Dependent Coding Techniques*, U.S. Patent 3,694,813, September 26, 1972.

43. J. P. McCarthy, *Automatic File Compression*, International Computing Conference 1973, 1974, 511-516.

44. G. N. N. Martin, *Range Encoding: an Algorithm for Removing Redundancy from a Digitized Message*, Video and Data Recording Conference, Southampton, July, 1979.

45. G. A. Miller, *The Magical number seven, plus or minus two: Some limits to our capacity for Processing Information*, The Psychological Review, *63* (56), 81-97

46. V. S. Miller and M. N. Wegman, *Variations on a theme by Ziv and Lempel*, in Combinatorial Algorithms on Words, edited by Apostolico and Galil, Springer-Verlag, 1985.

47. J. H. Mommens and J. Raviv, *Coding for Data Compaction*, IBM Research Report RC 5150, T. J. Watson Research Center, Yorktown Heights, NY, November 26, 1974.
48. H. Morita and S. Arimoto, *SECT-A Coding Technique for Black/White Graphics*, IEEE Trans. Inform. Theory, *IT-29* (83), 559-570
49. L. T. N. M. Mulder and D. L. Cohn, *An Adaptive Noiseless Variable-to-Fixed Encoding Algorithm for Discrete Ergodic Sources*, Proc. 15th Allerton Conference on Circuits and System Theory, 1977, 972-981.
50. J. Mulford, *Data Compression Techniques for Economic Processing of Large Commercial Files*, ACM Symp. on Info. Storage and Retrieval, 1971, 207-215.
51. R. Pasco, *Source coding algorithms for fast data compression*, PhD dissertation, Dept. EE, Stanford Univ. Stanford, Ca., May 1976.
52. M. Pechura, *File Archival Techniques Using Data Compression*, Comm. ACM, *25* (82), 605-609
53. Y. Perl, M. R. Garey and S. Even, *Efficient Generation of Optimal Prefix Code: Equiprobable Words Using Unequal Cost Letters*, JACM, *22* (75), 202-214
54. J. Pike, *Text Compression using a 4-bit coding scheme*, Computer Journal, *24* (81), 324-330
55. H. K. Reghbati, *An Overview of Data Compression Techniques*, IEEE Computer, *14* (81), 71-75
56. J. Rissanen, *Generalized Kraft-inequality and arithmetic coding*, IBM J. Res. Dev., *20* (76), 198-203
57. J. Rissanen, *Minimax codes for finite alphabets*, IEEE Trans. Inform. Theory, *IT-24* (78), 389-392
58. J. Rissanen, *Arithmetic codings as number representations*, Acta Pol. Scand. Math. *31* (79), 44-51
59. J. Rissanen and G. Langdon, Jr., *Arithmetic coding*, IBM J. Res. Dev., *23* (79), 149-162
60. J. Rissanen and G. Langdon, Jr., *Universal Modeling and Coding*, IEEE Trans. Inform. Theory, *IT-27* (81), 12-23
61. M. Rodeh, V. Pratt and S. Even, *Linear Algorithm for Data Compression via String Matching*, J. ACM, *28* (81), 16-24
62. F. Rubin, *Experiments in Text File Compression*, Comm. ACM, *19* (76), 617-623
63. F. Rubin, *Arithmetic Stream Coding Using Fixed Precision Registers*, IEEE Trans. Inform. Theory, *IT-25* (79), 672-675
64. S. S. Ruth and P. J. Kreatzer, *Data Compression for Large Business Files*, Datamation, *18* (72), 62-66
65. J. Schalkwijk, *An Algorithm for Source Coding*, IEEE Trans. Inform. Theory, *IT-18* (72), 395
66. E. J. Schuegraf and H. S. Heaps, *A Comparison of Algorithms for Data Base Compression by use of Fragments as Language Elements*, Inform. Stor. Retr. *10* (74), 309-319
67. C. E. Shannon, *A Mathematical Theory of Communication*, Bell System Tech. J., *27* (48), 379-423
68. C. E. Shannon, *Prediction and Entropy of Printed English*, Bell System Tech. J., *30* (51), 50-64
69. D. D. Sleator and R. E. Tarjan, *Amortized Efficiency of List Update and Paging Rules*, Bell Labs Technical Memorandum, 1983.
70. R. J. Solomonoff, *A Formal Theory of Inductive Inference*, Info. Control, *7* (64), 224
71. James A. Storer and T. G. Szymanski, *Data Compression via Textual Substitution*, J. ACM, *29* (82), 928-951
72. S. J. P. Todd, G. G. Langdon, Jr., and G. N. N. Martin, *A General Fixed-rate Arithmetic Coding Method for Constrained Channels*, IBM J. of Res. and Dev., *27* (83), 94-106
73. H. E. White, *Printed English Compression by Dictionary Encoding*, Proc. IEEE, *55* (67), 390-396
74. E. J. Yannakoudakis, P. Goyal and J. A. Huggill, *Generation and Use of Text Fragments for Data Compression*, Inf. Processing and Management, vol. 18, no. 1, 1982, 15-21.
75. J. Ziv, *Coding Theorems for Individual Sequences*, IEEE Trans. Inform. Theory, *IT-24* (78), 405-412
76. J. Ziv and A. Lempel, *A Universal Algorithm for Sequential Data Compression*, IEEE Trans. Inform. Theory, *IT-23* (77), 337-343
77. J. Ziv and A. Lempel, *Compression of Individual Sequences via Variable-Rate Coding*, IEEE Trans. Inform. Theory, *IT-24* (78), 530-536

Department of Mathematics, IBM Thomas J. Watson Research Center, Yorktown Heights, NY 10598

Proceedings of Symposia in Applied Mathematics
Volume 34, 1986

APPLICATION OF CATEGORY THEORY OF STRUCTURAL SETS TO MODELING OF INFORMATION BASES OF SYSTEMS

Augustin A. Tuzhilin

Contents

INTRODUCTION

In this paper, we develop, as a first approach, a methodology for mathematical modeling of information systems. The necessity for such a methodology is widely recognized. We construct a mathematical model for the information base of a system applying general methods of mathematical modeling known as structural approach (in N. Bourbaki sense, see [1]). The information base of a system describes the data structure for the entire system and, therefore, should be the main structure of the information system. The information system, as we view it, should contain the information base for the external world and the information base of a system storing and manipulating the external data. It seems that there exists the universal information system being capable of storing and manipulating any kind of information. Computers seem to be an approach to the universal information system. Therefore, the analysis of information bases of systems is the first step to the analysis of general information systems. In this paper, we do not consider processes in a system, which shold be described by the algebraic structure on the information base of a system more complex than the universal algebra. We feel that further applications of the developed methods to real information systems should justify the inclusion of additional structures into their models.

In this paper, we describe the extension of the N. Bourbaki theory of structural sets developed by us for the purposes of mathematical modeling in general [2]. We developed canonical representations of structures, which permitted us to introduce canonical morphisms and, therefore, categories for any given type of structures. This technique allows us to use the convenient language and results of the Theory of Categories for modeling purposes. These results are discussed in section 2. In section 1, we consider an example—a simplified hospital system. By developing a mathematical model for this system, we, first, demonstrate our general technique of mathematical modeling, and, second, we justify the introduction of the structure "the information base of a system." The description of the simplified hospital system structure and the general constructions of the structural sets theory use additional properties (axioms known as transferable relations) of structures. In section 3, we develop the technique for the construction of a language based on primitive sets (called basic sets in the theory of structural sets). We consider finite primitive sets and call this language "a universal query language on finite primitive sets." In section 3, we construct a universal query language for a given primitive set. We introduce an equality relation and logical operations, which allow us to define theorems and introduce quantifiers and a relation "belong" in the language. Many results of the set theory may be obtained in this language.

In section 4, we describe the general model for a system constructed by us in previous papers (see, for example, [2]) and discuss relations between systems and information bases of systems.

In sections 5 and 6, we consider specific examples of information bases of systems. In section 5, we consider functional information bases, formulate the composition problem for this type of bases and give the solution for the problem. We construct specific functional equations associated with a family of functional information bases and a scheme of composition and analyze the algebraic structure of and solutions for these equations.

In section 6, we develop further the theory of specific information bases called "spaces of (available) states." We describe several constructions in the theory in terms of category theory and develop convenient techniques for describing such bases. We apply results obtained to modeling computers and their languages. Main results of this section are published in [3] and [4].

1. INFORMATION BASE OF A SYSTEM

To justify the introduction of a general model for the information base of a system, we are going to consider, first, an example of a simplified hospital system. By analyzing the mathematical model designed for this system and particularly the information required for describing this system, we introduce a general model for the information base of a system. The mathematical model designed demonstrates the methodology of the theory of structural sets developed for mathematical modeling in general. We chose a simplified hospital system for modeling because, first, it contains some important features of the information bases of systems we are going to formalize; and, second, it is very far from mathematical practice and, so, needs its own mathematical methods for analysis. On the other hand, it is not difficult to realize some processes in such systems.

1.1 Simplified Hospital

Our hospital, for example, consists of two rooms (departments) r_1 and r_2. The first room, r_1, cures the diseases dS_{11}, dS_{12} and the second room, r_2, cures the diseases dS_{21}, dS_{22}. All these diseases are supposed to be distinct. We suppose that there are one bed, b_{11}, and two categories of specialized equipment, e_{11}, e_{12} in the first room, and two beds, b_{21}, b_{22}, and one category of specialized equipment, e_{21}, in the second room. All these categories, e_{11}, e_{12}, e_{21}, are distinct. It is supposed that category e_{11} cures the disease dS_{11}, e_{12} cures the disease dS_{11}, e_{21} cures, of course, all diseases of the second room, i.e., dS_{21} and dS_{22}. In the second room, r_2, it is supposed that the diseases dS_{21} and dS_{22} are cured on the bed b_{21} and the disease dS_{22} is cured on the bed b_{22}. Two doctors, d_1 and d_2, are supposed to work for the hospital. The first doctor, d_1, cures the diseases dS_{11} and dS_{22} on the bed b_{22}. The second doctor, d_2, cures the diseases dS_{11}, dS_{12}, dS_{21}, dS_{22}, and the disease dS_{22} he cures only on the bed b_{21}.

Let's suppose that at the given moment two patients, p_{H1} and p_{H2}, are cured by the hospital; one patient, p_{I1}, wants to be hospitalized, and one patient, p_{O1}, leaves the hospital. It is supposed that the patient p_{H1} has the diseases dS_{11}, dS_{12}, and dS_{22}, and is treated now for the diseases dS_{11} and dS_{12} on the bed b_{11} by the doctors d_1 (for dS_{11} disease) and d_2 (for dS_{12} disease); he is also assigned to the categories of specialized equipment e_{11} (for the disease dS_{11}) and e_{12} (for the disease dS_{12}). The second patient p_{H2} is treated for the disease dS_{22} on the bed b_{21} by the doctor d_2 and is assigned to e_{21}. The patient p_{O1} was treated for the disease

dS_{22} on the bed b_{22} by the doctor d_1 and was assigned to e_{21}. The patient p_{I1} has the diseases dS_{11}, dS_{22} and must be assigned to a room, bed, doctor, and category of specialized equipment. It is supposed that any patient can be assigned without changing his treatment except with healing his diseases. Hence, the processes in this system consist of changing the status of patients compatible with the hospital status.

I described a very simple system. In the next section, I will define the general hospital system of this kind. The described system may be visualized as in Figure 1.

Let us try to describe this system in a formal way. We should introduce six primitive sets:

$R = \{r_1, r_2\}$ is a set of rooms (departments) in the hospital.

$B = \{b_{11}, b_{21}, b_{22}\}$ is a set of beds in the hospital.

$\Sigma = \{e_{11}, e_{12}, e_{21}\}$ is a set of categories of specialized equipment in the hospital.

$DS = \{dS_{11}, dS_{12}, dS_{21}, dS_{22}\}$ is a set of diseases cured in the hospital.

$D = \{d_1, d_2\}$ is a set of doctors working for the hospital.

$P = \{p_{H1}, p_{H2}, p_{O1}, p_{I1}\}$ is a set of patients which should be cured in the hospital.

But these six sets are related. These relations will be described by specific elements of complex sets constructed from primitive sets by applying only operation of cartesian products and constructing all subsets (boolean operation). Namely, beds are partitioned among rooms. In general, a partition of any set X is a family $(X_i)_{i \in I}$ of sets so that $X_i \cap X_j = \phi$ for $i \neq j$ and $X = \cup_{i \in I} X_i$, or, in other words,

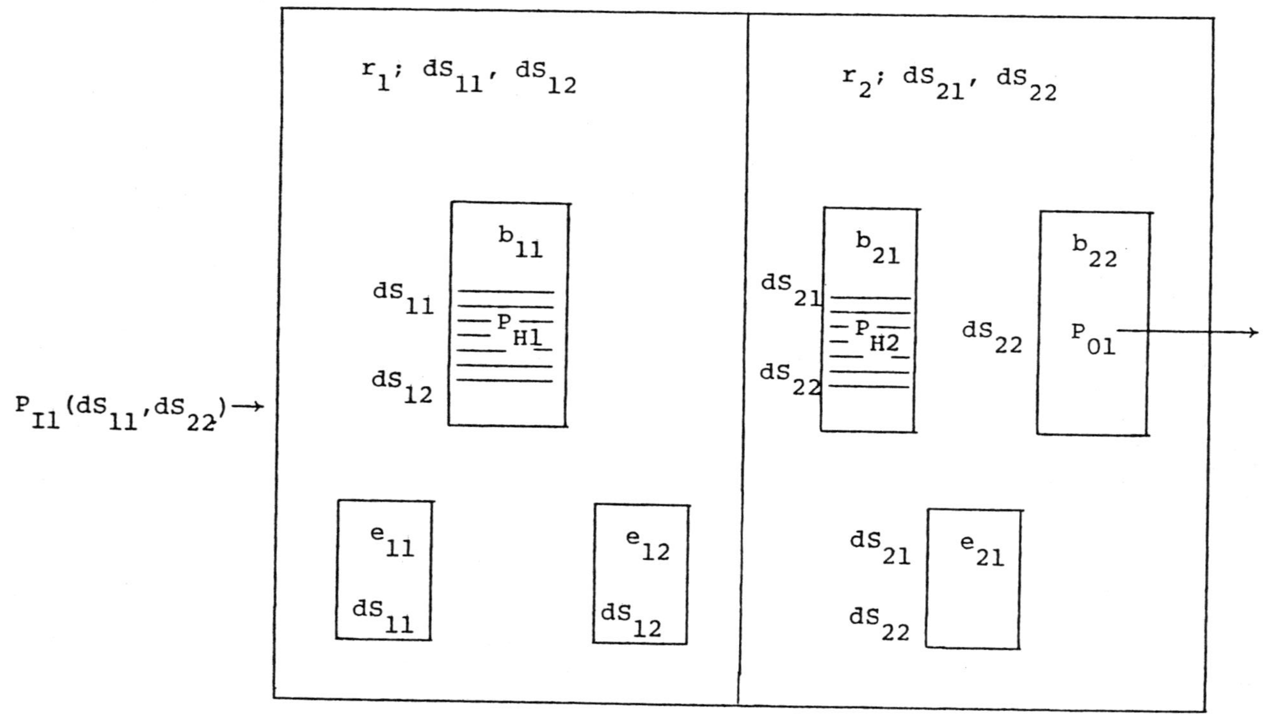

$p_{H1}(dS_{11}(d_1,e_{11}),dS_{12}(d_2,e_{12}),dS_{22})$, $p_{H2}(dS_{22}(d_2,e_{21})p_{01}(dS_{22}(d_1,e_{21}))$,

$d_1(dS_{11},dS_{22}(b_{22}))$, $d_2(dS_{11},dS_{12},dS_{21},dS_{22}(b_{21}))$.

Figure 1

there is a mapping $P_X^I: I \to \beta(X)$, where $\beta(X)$ is the set of all subsets of X (boolean of X), so that $P_X^I(i) = X_i$. The graph of the mapping P_X^I is an element $P_X^I \in \beta(X \times \beta(X))$. P_X^I is a functional graph and is an injection, that is, $P_X^I(i) \cap P_X^I(j) = \phi$ for $i \neq j$, and $\cup_{i \in I} P_X^I(i) = X$. For our hospital we have

$$P_B^R \in \beta(R \times \beta(B)) \tag{1.1}$$

where $P_B^R(r_1) = \{b_{11}\}$, $P_B^R(r_2) = \{b_{21}, b_{22}\}$. Diseases are distributed among beds. We define a distribution F_X^I of a set X among elements of a set I as a family $(X_i)_{i \in I}$, or $F_X^I \in \beta(I \times \beta(X))$ so that F_X^I is a mapping of I into $\beta(X)$; $F_X^I(i) = X_i$ and $X = \cup_{i \in I} F_X^I(i)$. Therefore, for the hospital we have

$$F_{DS}^B \in \beta(B \times \beta(DS)) \tag{1.2}$$

so that $F_{DS}^B(b_{11}) = \{dS_{11}, dS_{12}\}$, $F_{DS}^B(b_{21}) = \{dS_{21}, dS_{22}\}$, $F_{DS}^B(b_{22}) = \{dS_{22}\}$. We suppose that if b_1 and b_2 are beds in different rooms, that is, $b_1 \in P_B^R(r_1)$, $b_2 \in P_B^R(r_2)$, and $r_1 \neq r_2$, then diseases $F_{DS}^B(b_1)$ and $F_{DS}^B(b_2)$ are different, that is, $F_{DS}^B(b_1) \cap F_{DS}^B(b_2) = \phi$. Therefore, the distribution

$$P_{DS}^R = F_{DS}^B \circ P_B^R \tag{1.3}$$

is a partition of the diseases DS among the rooms R. In our case $P_{DS}^R(r_1) = \{dS_{11}, dS_{12}\}$, $P_{DS}^R(r_2) = \{dS_{21}, dS_{22}\}$. Diseases

DS are also distributed among categories of specialized equipment, so that all diseases for every $e \in \Sigma$ are treated for only in one room. Therefore, we have

$$F^{\Sigma}_{DS} \in \beta(\Sigma \times \beta(DS)) \tag{1.4}$$

and if $F^{\Sigma}_{DS}(e)$ is a set of diseases assigned to the equipment $e \in \Sigma$, then there is a room $r \in R$ so that $F^{\Sigma}_{DS}(e) \subset P^{R}_{DS}(r)$. Hence the distribution

$$P^{R}_{\Sigma} = F^{-1}{}^{\Sigma}_{DS} \circ P^{R}_{DS} = F^{-1}{}^{\Sigma}_{DS} \circ F^{B}_{DS} \circ P^{R}_{B} \tag{1.5}$$

is a partition of equipment Σ among the rooms R (F^{-1} denotes the inverse of the graph F). In our case, we have $F^{\Sigma}_{DS}(e_{11}) = \{dS_{11}\}$, $F^{\Sigma}_{DS}(e_{12}) = \{dS_{12}\}$, $F^{\Sigma}_{DS}(e_{21}) = \{dS_{21}, dS_{22}\}$, $P^{R}_{\Sigma}(r_1) = \{e_{11}, e_{12}\}$, $P^{R}_{\Sigma}(r_2) = \{e_{21}\}$.

Diseases are also distributed among doctors, so we have

$$F^{D}_{DS} \in \beta(D \times \beta(DS)) \quad . \tag{1.6}$$

But we suppose that disease dS being treated by doctor d, that is, $dS \in F^{D}_{DS}(d)$, is treated for by him on some beds $F^{-1}{}^{B}_{DS}(dS)$ available for the disease. It means that for any pair (d,dS), where dS is a disease cured by the doctor d, that is, $dS \in F^{D}_{DS}(d)$, there are beds on which the doctor d cures this disease. So we have the set of pairs

$$P(D,DS) \in \beta(D \times DS) \tag{1.7}$$

so that

$$(d,dS) \in P(D,DS) \iff dS \in F^{D}_{DS}(d) \quad , \tag{1.8}$$

and the distributions

$$F_B^{P(D,DS)} \in ((D \times DS) \times \beta(B)) \tag{1.9}$$

so that

$$pr_1 F_B^{P(D,DS)} = P(D,DS) \quad ,$$

$$F_B^{P(D,DS)}(d,dS) \subset \overset{-1}{F}{}_{DS}^{B}(ds) \tag{1.10}$$

In our case

$$F_{DS}^{D}(d_1) = \{dS_{11}, dS_{22}\} \quad ,$$

$$F_{DS}^{D}(d_2) = \{dS_{11}, dS_{12}, dS_{21}, dS_{22}\} \quad ,$$

$$F_B^{P(D,DS)}(d_1, dS_{11}) = F_B^{P(D,DS)}\left(d_2, \begin{Bmatrix} dS_{11} \\ dS_{12} \end{Bmatrix}\right) = \{b_{11}\} \quad ,$$

$$F_B^{P(D,DS)}(d_1, dS_{22}) = F_B^{P(D,DS)}\left(d_2, \begin{Bmatrix} dS_{21} \\ dS_{22} \end{Bmatrix}\right) = \{b_{21}\} \quad .$$

Primitive sets R, B, Σ, DS, D and elements P_B^R, F_{DS}^B, F_{DS}^{Σ}, F_{DS}^D, $F_B^{P(D,DS)}$ which hold the conditions (1.1) - (1.10) describe the hospital. Therefore, we introduce the object

$$H = (R,B,\Sigma,DS,D;\ P_B^R, F_{DS}^B, F_{DS}^{\Sigma}, F_{DS}^D, F_B^{P(D,DS)}) \tag{1.11}$$

which will be called a hospital structure. We can also say that the sets R, B, Σ, DS, and D are endowed with structures P_B^R, F_{DS}^B, F_{DS}^{Σ}, F_{DS}^D, and $F_B^{P(D,DS)}$.

To analyze further the relations between primitive sets R, B, Σ, DS, D, and P, we should consider our hospital as a system. A hospital as an object is characterized by the structure H defined in (1.11). Patients which are treated for in the hospital characterize the current state of the hospital. Patients waiting for treatment characterize the state of the input to the hospital, and patients leaving the hospital characterize the state of the output of the hospital. Therefore, we can represent the hospital as a graph with one input, I, one element, H, and one output, O:

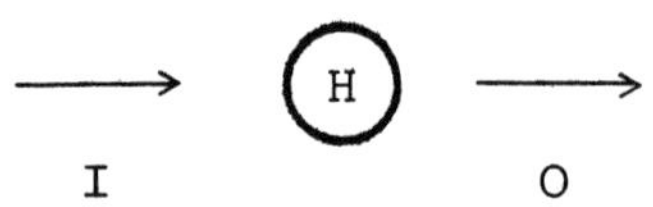

The element H is described by the structure (1.11). In our case, we do not assign any structures to input I and output O. But let us define spaces of states for H, I, and O. First, we define a patient structure for a set of patients P. A set of patients P will be characterized by the set of diseases T^P_{DS} as a mapping $T^P_{DS}: P \to \beta(DS)$ so that for every patient $p \in P$ we have the set of diseases $T^P_{DS}(p) \subset DS$ (we do not require the condition $\cup_{p \in P} T^P_{DS}(p) = DS$ to hold). The patient structure is the object $P^S = (P, DS; T^P_{DS})$.

Let us now define the space of states of the hospital H as an element of the system. For a given patient structure P^S and hospital structure H, we introduce the following mappings:

A mapping

$$f^P_B: P \to B \tag{1.12}$$

is an injection; it assigns the bed $f^P_B(p) \in B$ for every patient $p \in P$ (we assume that the number of patients does not exceed the number of beds). In our case, we have patient p_{H1} with diseases ds_{11}, ds_{12}, and ds_{22}, and p_{H2} with disease ds_{22}, that is $P_H = \{p_{H1}, p_{H2}\}$. We have

$$f_B^{P_H}(p_{H1}) = b_{11} \quad , \quad f_B^{P_H}(p_{H2}) = b_{12} \quad .$$

A mapping

$$f_{DS}^P : P \to \beta(DS) \qquad (f_{DS}^P(p) \subset T_{DS}^P(p)) \tag{1.13}$$

assigns to every patient $p \in P$ a set of diseases $f_{DS}^P(p) \subset DS$ that will be treated for on his/her bed $f_B^P(p)$. So, since $F_{DS}^B(b)$ is a set of diseases which are treated for on the bed b,

$$f_{DS}^P(p) \subset F_{DS}^B(f_B^P(p)) \tag{1.14}$$

In our case, $f_{DS}^{P_H}(p_{H1}) = \{ds_{11}, ds_{12}\}$, $f_{ds}^{P_H}(p_{H2}) = \{ds_{22}\}$.

A mapping

$$f_\Sigma^P : P \to \beta(\Sigma) \tag{1.15}$$

assigns to every patient $p \in P$ a set of equipment $f_\Sigma^P(p) \subset \Sigma$ that may cure the diseases $f_{DS}^P(p)$ of that patient. So,

$$f_{DS}^P(p) \subset F_{DS}^\Sigma(f_\Sigma^P(p)) \quad . \tag{1.16}$$

In our case, $f_\Sigma^{P_H}(p_{H1}) = \{e_{11}, e_{12}\}$, $f_\Sigma^{P_H}(p_{H2}) = \{e_{21}\}$.

A mapping

$$f_D^P : P \to \beta(D) \tag{1.17}$$

assigns to every patient $p \in P$ a set of doctors $f_D^P(p) \in D$ thay may cure the diseases $f_{DS}^P(p)$ of the patient. So,

$$f^P_{DS}(p) \subset F^D_{DS}(f^P_D(p)) \tag{1.18}$$

In our case, $f_D^{P_H}(p_{H1}) = \{d_1,d_2\}$, $f_D^{P_H}(p_{H2}) = \{d_2\}$.

The array of mappings $A^P = (f^P_B, f^P_{DS}, f^P_\Sigma, f^P_D)$, which holds the conditions (1.12) - (1.18) is called an assigning array. A pair (P^S,A^P), where P^S is a patient structure and A^P is an assigning array, is called an available state of the hospital H. The space of states of the hospital H is the set of all available states of H. In our case, the array $A^{P_H} = (f_B^{P_H}, f_{DS}^{P_H}, f_\Sigma^{P_H}, f_D^{P_H})$ is an assigning array. Now, we define the spaces of states for input I and output O. The space of states of input I consists of all patients' structures P^S. The space of states of output O consists of all pairs (P^S,A^P), that is, the same spaces as for the hospital.

To define the space of availabile states for the whole system, we define, first, the partial operation on pairs (P^S,A^P).

If for two pairs (P_1^S,A^{P_1}) and (P_2^S,A^{P_2}) such that P_1 and P_2 are different, that is $P_1 \cap P_2 = \phi$, there exists an array $(f_B^{P_1+P_2}, f_{DS}^{P_1+P_2}, f_\Sigma^{P_1+P_2}, f_D^{P_1+P_2})$ such that a mapping

$$f_B^{P_1+P_2}: P_1 \cup P_2 \to B: p \to f_B^{P_1+P_2}(p) = \begin{cases} f_B^{P_1}(p) \text{ for } p \in P_1 \\ f_B^{P_2}(p) \text{ for } p \in P_2 \end{cases}$$

is an injection;

$$f_X^{P_1+P_2} : P_1 \cup P_2 \to \beta(X) : p \to f_X^{P_1+P_2}(p) = \begin{cases} f_X^{P_1}(p) & \text{for } p \in P_1 \\ f_X^{P_2}(p) & \text{for } p \in P_2 \end{cases}$$

for $X = DS, \Sigma, D$, and for $p \in P_1 \cup P_2$

$$f_X^{P_1+P_2}(p) \subset F_X^Y \left(f_Y^{P_1+P_2}(p)\right)$$

for $Y = B, \Sigma, D$; then we denote this array as

$$A^{P_1+P_2} = \left(f_B^{P_1+P_2}, f_{DS}^{P_1+P_2}, f_\Sigma^{P_1+P_2}, f_D^{P_1+P_2}\right) .$$

The patient structure $(P_1 \cup P_2,\ DS;\ T_{DS}^{P_1+P_2})$ where

$$T_{DS}^{P_1+P_2} : P_1 \cup P_2 \to \beta(DS) : p \to T_{DS}^{P_1+P_2}(p) = \begin{cases} T_{DS}^{P_1}(p) & \text{for } p \in P_1 \\ T_{DS}^{P_2}(p) & \text{for } p \in P_2 \end{cases}$$

will be denoted $P_1^S + P_2^S$. So, we may define the partial operation on pairs (P^S, A^P) as

$$(P_1^S, A^{P_1}) + (P_2^S, A^{P_2}) = (P_1^S + P_2^S, A^{P_1+P_2}) .$$

We will define the space of available states of the system by assuming that there is no queue for the treatment. The space of available states S^H of the system consists of the objects $(P_I^S,(P_H^S,A^{P_H}),(P_O^S,A^{P_O}))$ so that there exists an assigning array A^{P_I} such that the sums

$$(P_I^S,A^{P_I}) + (P_H^S,A^{P_H}) \quad \text{and} \quad (P_H^S,A^{P_H}) + (P_O^S,A^{P_O}) \qquad (1.19)$$

exist.

For A^{P_O}, where $P_O = \{p_{O1}\}$, $T_{DS}^{P_O}(p_{O1}) = \{dS_{22}\}$, we have

$$f_B^{P_O}\colon \{p_{O1}\} \to \{b_{11},b_{21},b_{22}\} \ ; \quad f_B^{P_O}(p_{O1}) = b_2$$

$$f_{DS}^{P_O}\colon \{p_{O1}\} \to \beta(DS) \ ; \quad f_B^{P_O}(p_{O1}) = \{dS_{22}\}$$

$$f_\Sigma^{P_O}\colon \{p_{O1}\} \to \beta(\{e_{11},e_{12},e_{21}\}) \ ; \quad f_\Sigma^{P_O}(p_{O1}) = \{e_{21}\}$$

$$f_D^{P_O}\colon \{p_{O1}\} \to \beta(\{d_1,d_2\}) \ ; \quad f_D^{P_O}(p_{O1}) = \{d_1\} \quad .$$

Array A^{P_O} is also an assigning array. The state

$$(P_I^S,(P_H^S,A^{P_H}),(P_O^S,A^{P_O}))$$

is an available state since for the assigning array A^{P_I}, where $P_I = \{p_{I1}\}$, $T_{DS}^{P_H}(p_{I1}) = \{dS_{11},dS_{22}\}$, we have

$$f_B^{P_I}\colon \{p_{I1}\} \to \{b_{11},b_{21},b_{22}\} \ ; \quad f_B^{P_I}(p_{I1}) = b_{22}$$

$$f_{DS}^{P_I}\colon \{p_{I1}\} \to \beta(DS) \ ; \quad f_{DS}^{P_I}(p_{I1}) = \{dS_{22}\}$$

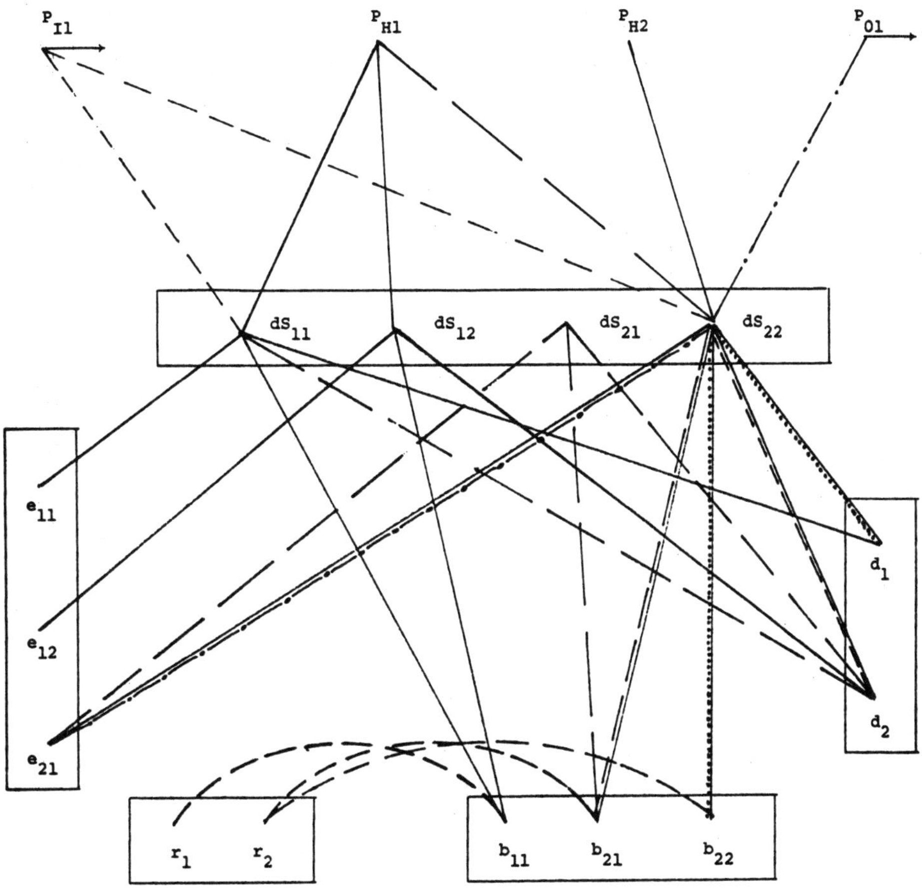

p_{I1}
p_{H1}
p_{H2}
p_{01}
ds_{11}
ds_{12}
ds_{21}
ds_{22}
e_{11}
e_{12}
e_{21}
d_1
d_2
r_1
r_2
b_{11}
b_{21}
b_{22}

Figure 2

$$f_{\Sigma}^{P_I}: \{p_{I1}\} \to \beta(\{e_{11}, e_{12}, e_{21}\}) \ ; \quad f_{\Sigma}^{P_I}(p_{I1}) = \{e_{21}\}$$

$$f_{D}^{P_I}: \{p_{I1}\} \to \beta(\{d_1, d_2\}) \ ; \quad f_{D}^{P_I}(p_{I1}) = \{d_1\}$$

the sums $(P_I^S, A^{P_I}) + (P_H^S, A^{P_H})$ and $(P_H^S, A^{P_H}) + (P^{S_O}, A^{P_O})$ exist. The whole information about the system can be described by the following pair

$$IB_{SYS} = (\Gamma, S^H) \tag{1.20}$$

where Γ is an alement base defined by the graph with one input I, one element H, and one output O. The structure H defined in (1.11) is assigned to the element H. S^H is the space of available states for the element base Γ. This pair will be called an information base of a system (for the element base Γ). In our case, the information base of the system is represented in Figure 2, where different lines describe different conditions from (1.1) to (1.19).

1.2 A General Mathematical Model for the Information Base of a System

In the previous point, we constructed a mathematical model for the simplified version of a hospital. A hospital, as an element of the system, constructed was described by the structures in (1.11). In this structures, we assumed that all sets are specific sets. But we can now assume that in (1.11) all sets—primitive and complex—are arbitrary sets, which hold the properties of (1.1) through (1.18). Therefore, in (1.11) we described a general hospital of the type represented by a general structure in (1.11) where all letters are variables. We can consider relations (1.19) and (1.20) in the same way, and therefore the relation (1.20) describes the information base of a general hospital of a given type.

In this point, we are going to generalize the construction developed in the previous point and to construct a general model for information base. The first object we should introduce is an element base (equipment) which supports the information. In the precious point it was a graph with one element, one input, and one output. We will restrict ourselves here with the case of a "finite equipment," which consists of inputs, elements (parts), connections between elements, and outputs. This type of "equipment" describes many practical situations, computers for example. Therefore, the element base should be described by the following primitive sets: A is a set of elements, C is a set of connections between elements, I is a set of inputs, O is a set of outputs. We assume that every connection $c \in C$ connects two elements: the initial and the final element. We assume also that every input $i \in I$ enters into and every output $o \in O$ leaves one of the elements from A. Therefore, we should have two functions $\partial_0 : C \cup O \to A$, $\partial_1 : C \cup I \to A$, which assign the initial or the final elements to a connection, output, and input. Therefore, the element base B is a structure

$$B\ (A, C, I, O;\ \partial_0, \partial_1)$$

where $\partial_0 \in \beta((C \cup O) \times A)$, $\partial_1 \in \beta((C \cup I) \times A)$ (if we don't want to use the union operation, we can define partial functions ∂_0 and ∂_1). An element x of the set $A \cup C \cup I \cup O$

will be called a base element. We will write $x \in B$, by abusing language. Properties of base elements can be described by structures. Therefore, we should have a family $(P_x)_{x \in B}$ of structures.* If we do not want to assign a structure to an element of B, then we assign the empty set (it is an object in the category of sets) to this element. Therefore, equipment is described by two objects

$$EQ = (B, (P_x)_{x \in B}) = ((A,C,I,O; \partial_0,\partial_1), (P_x)_{x \in B}) \quad .$$

Here sets A, C, I, and O give "labels" to the elements of equipment, the structure P_x for $x \in A \cup C \cup I \cup O$ describes properties of the element x.

The next objects to be introduced are spaces of states of the elements $x \in B$. In the previous point, spaces of states of the input, the element H, and the output were sets of structures. Therefore, we should have a family $(K_x)_{x \in B}$ of sets K_x of structures of given types, that is, every element of K_x is a structure of a given type (an object of some category, see below). The state of the equipment, now, is a mapping s, which puts every element x from B into one of the possible states for this element, that is,

$$s: x \to s(x) \qquad (x \in B, \ s(x) \in K_x)$$

In other words, a state of the equipment is a family $(s(x))_{x \in B}$ and $s(x) \in K_x$. We denote all possible states by

---

* We will show later that structures of a given type are objects of some categories. Therefore, a family of structures is a meaningful object.

$\prod_{x \in B} K_x$. The space of available states S_{EQ} is a subset of all possible states, that is,

$$S_{EQ} \subset \prod_{x \in B} K_x \quad .$$

The information base IB_{SYS} in general is a pair

$$IB_{SYS} = (EQ, S_{EQ}) \quad .$$

We introduced very complex mathematical object for the information base. To describe this object completely we need some techniques from the general theory of categories of structural sets. For example, we need techniques for handling families of structures. Some aspects of the general theory of categories of structural sets are discussed in the next section.

2. THEORY OF CATEGORIES OF STRUCTURAL SETS

2.1 Echalons

The goal of this point is to introduce a certain class of sets which will be used in constructing structural sets.

Let S be a finite set of distinct signs $x_1, x_2, \ldots, x_n$, P, P^-, X, X^-. The set of all finite words in S will be denoted by S*, i.e., $S^* = \cup_{n \geq 0} S^n$. The signs $x_1, x_2, \ldots, x_n$ will be called *letters*.

Definition 2.1. Sequence $(T_i)_{1 \leq i \leq n}$ of words of S* is called a signed significant sequence and its elements are called a covariant or contravariant echalon type if and only if for every T_i one of the following statements is correct.

1. T_i is a letter. In this case T_i is called a covariant echalon type.
2. There is a covariant (contravariant) echalon type T_j for $j < i$ such that $T_i = PT_j$ or $T_i = P^- T_j$. The word PT_j is called a covariant (contravariant) and $P^- T_j$ a contravariant (covariant) echalon type.
3. There are two covariant (contravariant) echalon types T_{i_1} and T_{i_2} for $i_1 < i$ and $i_2 < i$ such that $T_i = XT_{i_1}T_{i_2}$ $(T_i = X^- T_{i_1}T_{i_2})$. In this case T_i is called a covariant (contravariant) echalon type.

An element of any signed significant sequence is also called a *signed echalon type*. The signed echalon type T in the alphabet S will be often denoted by $T(x_1, x_2, \ldots, x_n)$. Let $E_1, \ldots, E_n$ be n sets. For every signed ecahlon type $T(x_1, \ldots, x_n)$ we associate the set $T(E_1, \ldots, E_n)$ by the following

1. If $T(x_1, \ldots, x_n) = x_i$ then $T(E_1, \ldots, E_n) = E_i$.
2. If $T(x_1, \ldots, x_n) = PU(x_1, \ldots, x_n)$ or $P^- U(x_1, \ldots, x_n)$, then $T(E_1, \ldots, X_n) = \beta(U(E_1, \ldots, E_n))$.
3. If $T(x_1, \ldots, x_n) = XU(x_1, \ldots, x_n)U_2(x_1, \ldots, x_n)$ or $X^- U(x_1, \ldots, x_n)U_2(x_1, \ldots, x_n)$, then $T(E_1, \ldots, E_n) = U_1(E_1, \ldots, E_n) \times U_2(E_1, \ldots, E_n)$.

Here $\beta(E)$ is the set of all subsets of the set E (boolean of E). The set $T(E_1, \ldots, E_n)$ is called the *realization of the signed echalon type* $T(x_1, \ldots, x_n)$, *or covariant and contravariant echalon* (in accordance with covariancy and contravariancy of T), *on sets* $E_1, \ldots, E_n$. From the above construction it follows

Theorem 2.1. *For every covariant or contravariant echalon* $T(E_1,\ldots,E_n)$ *there exist* p *echalons* $T_1(E_1,\ldots,E_n)$, $\ldots,T_p(E_1,\ldots,E_n)$ *and a bijection of: covariant echalon* $T(E_1,\ldots,E_n)$ *into the covariant echalon*

$$\beta(T_1(E_1,\ldots,E_n)) \times \ldots \times \beta(T_p(E_1,\ldots,E_n)) \times E_{i_1} \times \ldots \times E_{i_k} ,$$

where $\beta(T_i(E_1,\ldots,E_n))$ *for* $1 \leq i \leq p$ *is a realization of covariant type* PT_i *for covariant* T_i *or* $P^- T_i$ *for contravariant* T_i *on* $E_1,\ldots,E_n$*; contravariant echalon* $T(E_1,\ldots,E_n)$ *into the contravariant echalon*

$$\beta(T_1(E_1,\ldots,E_n)) \times \ldots \times \beta(T_p(E_1,\ldots,E_n)) ,$$

where $\beta(T_i(E_1,\ldots,E_n))$ *for* $1 \leq i \leq p$ *is a realization of contravariant type* PT_i *for contravariant* T_i *or* $P^- T_i$ *for covariant* T_i *on* $E_1,\ldots,E_n$.

For a mapping $f\colon E \to E'$ we denote

$$f_\beta\colon \beta(E) \to \beta(E')\colon U \to f_\beta(U) = \{y\colon y = f(x) \text{ for some } x \in U)$$

$$f^\beta\colon \beta(E') \to \beta(E)\colon V \to f^\beta(V) = \{x\colon y = f(x) \text{ for some } y \in V\} .$$

The mappings f_β and f^β are called correspondingly the *canonical extension* and *inverse canonical extension of* f *into sets of subsets.*

For mappings $f\colon E \to E'$, $g\colon F \to F'$ we denote

$$f \times g\colon E \times F \to E' \times F'\colon (x,y) \to f \times g(x,y) = (f(x),g(y))$$

and call $f \times g$ the *canonical extension of* f *and* g *into products of sets.*

The set of mappings $f_1\colon E_1 \to E_1',\ldots, f_n\colon E_n \to E_n'$ will be denoted by $\bar{f}\colon \bar{E} \to \bar{E}'$.

Let $T(\overline{x}) = T(x_1,\ldots,x_n)$ be a signed echalon type and $T(\overline{E}) = T(E_1,\ldots,E_n)$ its realization on sets $E_1,\ldots,E_n$. We denote

1. If $T(\overline{x}) = x_i$, then T is covariant, $T(\overline{E}) = E_i$ and $\overline{f}_T = f_i: T(\overline{E}) \to T(\overline{E}')$.

2. If $T(\overline{x}) = Px_i$ (or $\overline{P}x_i$), then T is covariant (contravariant) $T(\overline{E}) = \beta(E_i)$ and

   $\overline{f}_T = (f_i)_\beta: T(\overline{E}) \to T(\overline{E}')$

   (or

   $\overline{f}^T = f_i^\beta: T(\overline{E}') \to T(\overline{E})$.)

3. If $T(\overline{x}) = Xx_1x_j$ then T is covariant $T(\overline{E}) = E_i \times E_j$ and

   $\overline{f}_T = f_i \times f_j: T(\overline{E}) \to T(\overline{E}')$.

4. If $T(\overline{x}) = \overline{X}\,\overline{P}x_i\overline{P}x_j$ then T is contravariant, $T(\overline{E}) = \beta(E_i) \times \beta(E_j)$ and

   $\overline{f}^T = f_1^\beta \times f_1^\beta: T(\overline{E}') \to T(\overline{E})$.

   Let $U(\overline{x})$ and $V(\overline{x})$ be signed echalon types consisting of m or less letters and for which the extensions

   $\overline{f}_u: U(\overline{E}) \to U(\overline{E}')$, $\overline{f}_v: V(\overline{E}) \to V(\overline{E})$

   for covariant $U(\overline{x})$ and $V(\overline{x})$; or

   $\overline{f}^u: U(\overline{E}') \to U(\overline{E})$, $\overline{f}^v: V(\overline{E}') \to V(\overline{E})$

   for contravariant $U(\overline{x})$ and $V(\overline{x})$ have beed defined. Then we define

5. If $T(\overline{x}) = PU(\overline{x})$, then for covariant (or contravariant) $U(\overline{x})$, $T(\overline{x})$ is covariant (contravariant), and

   $\overline{f}_T = (\overline{f}_u)_\beta: \beta(U(\overline{E})) \to \beta(U(\overline{E}'))$

(or

$$\overline{f}^{T} = (\overline{f}_{u})^{\beta}\colon \beta(U(\overline{E}')) \to \beta(U(\overline{E})) \quad .)$$

6. If $T(\overline{x}) = P^{-}U(\overline{x})$, then for covariant (contravariant) $U(\overline{x})$, $T(\overline{x})$ is contravariant (or covariant), and

$$\overline{f}^{T} = (\overline{f}_{u})^{\beta}\colon \beta(U(\overline{E}')) \to \beta(U(\overline{E}))$$

(or

$$\overline{f}_{T} = (\overline{f}^{u})^{\beta}\colon \beta(U(\overline{E})) \to \beta(U(\overline{E}')) \quad .)$$

7. If $T(\overline{x}) = XU(\overline{x})V(\overline{x})$, then $U(\overline{x})$, $V(\overline{x})$, $T(\overline{x})$ are covariant, and

$$\overline{f}_{T} = \overline{f}_{u} \times \overline{f}_{v}\colon U(\overline{E}) \times V(\overline{E}) \to U(\overline{E}') \times V(\overline{E}') \quad .$$

8. If $T(\overline{x}) = X^{-}U(\overline{x})V(\overline{x})$, then $U(\overline{x})$, $V(\overline{x})$, $T(\overline{x})$ are contravariant, and

$$\overline{f}^{T} = \overline{f}^{u} \times \overline{f}^{v}\colon U(\overline{E}') \times V(\overline{E}') \to U(\overline{E}) \times V(\overline{E}) \quad .$$

The mappings $\overline{f}^{T}$ or $\overline{f}_{T}$ are called the *signed canonical extensions of the mapping* $\overline{f}$ *associated with the signed echalon type* T. We see that for covariant (contravariant) type T a mapping $\overline{f}_{T}$ (or $\overline{f}^{T}$) is the mapping of $T(\overline{E})$ into $T(\overline{E}')$ (or of $T(\overline{E}')$ into $T(\overline{E})$). From the above definitions and constructions follow

Theorem 2.2. *If* $\overline{f}\colon \overline{E} \to \overline{E}'$ *and* $\overline{g}\colon \overline{E}' \to \overline{E}''$ *then*

$$(\overline{g} \circ \overline{f})_{T} = \overline{g}_{T} \circ \overline{f}_{T}\,, \qquad (\overline{g} \circ \overline{f})^{T} = \overline{f}^{T} \circ \overline{g}^{T} \quad ,$$

where $\overline{g} \circ \overline{f}$ *is a sequence* $g_1 \circ f_1\colon E_1 \to E''_1, \ldots, g_n \circ f_n\colon E_n \to E''_n$. *If for every* i, f_i *is a bijection of* E_i *into* E'_i *and* g_i *is the inverse bijection, then* $\overline{f}_{T}\colon T(\overline{E}) \to T(\overline{E}')$ *or* $\overline{f}^{T}\colon T(\overline{E}') \to T(\overline{E})$

is a bijection and $\overline{g}_T$ *or* $\overline{g}^T$ *is the corresponding inverse bijection.*

2.2 Categories of Structural Sets

The *typification on base sets* $\overline{x} = (x_1,\ldots,x_n)$ ($x_1,\ldots,x_n$ are distinct letters) *and auxillary sets* $\overline{A} = (A_1,\ldots,A_m)$ is a relation

$$\overline{U}^1 = (U_1^1,\ldots,U_p^1) \in A^1(\overline{x},\overline{A}) = \beta(A_1^1(\overline{x},\overline{A})) \times \ldots \times \beta(A_p^1(\overline{x},\overline{A})) \quad ,$$

$$\overline{U}_1 = (U_{11},\ldots,U_{1\ell}) \in A^1(\overline{x},\overline{A}) = \beta(A_{11}(\overline{x},\overline{A})) \times \ldots \times \beta(A_{1\ell}(\overline{x},\overline{A})) \quad ,$$

$$\overline{V} = (V_1,\ldots,V_k) \in x_{i_1} \times \ldots \times x_{i_k} \quad ,$$

where $A^1(\ldots)$ is a contravariant echalon, $A_1(\ldots)$ is a covariant echalon, and $U_1^1,\ldots,U_p^1$, $U_{11},\ldots,U_{1\ell}$, $V_1,\ldots,V_k$ are distinct letters. This relation will be denoted by $T(\overline{x},\overline{U}^1,\overline{U}_1,\overline{V},\overline{A})$. By $R(\overline{x},\overline{U}^1,\overline{U}_1,\overline{V})$ we denote a transportable relation (see [1]). A theory stronger than the set theory and having, as an addition axiom Σ, some typification $T(\overline{x},\overline{U}^1,\overline{U}_1,\overline{V},\overline{A})$ and some transportable relation $R(\overline{x},\overline{U}^1,\overline{U}_1,\overline{V})$ is called the *theory of species of structure* Σ. The term $(\overline{U}^1,\overline{U}_1,\overline{V})$ is called the *general species*.

Definition 2.2. If for sets $\overline{E}(E_1,\ldots,E_n)$, $\overline{P}^1 = (P^1,\ldots,P^p)$, $\overline{P}_1 = (P_1,\ldots,P_\ell)$, $\overline{V} = (V_1,\ldots,V_k)$ the relations $T(\overline{E},\overline{P}^1,\overline{P}_1,\overline{V},\overline{A})$, $R(\overline{E},\overline{P}^1,\overline{P}_1,\overline{V})$ are theorems, it is said the sets $E_1,\ldots,E_n$ are endowed with structure $(\overline{P}^1,\overline{P}_1,\overline{V})$ of the species Σ. Objects of the type $(E_1,\ldots,E_n;\ P^1,\ldots,P^p,\ P_1,\ldots,P_\ell,\ V_1,\ldots,V_k) = (\overline{E};\ \overline{P}^1,\overline{P}_1,\overline{V})$ will be called structural sets.

In the following by abusing the denotation we will denote by (see previous section) $\bar{f}_T: T(\bar{E},\bar{A}) \to T(\bar{E}',\bar{A})$ or $\bar{f}^T: T(\bar{E}',\bar{A}) \to T(\bar{E},\bar{A})$ the extensions of $\bar{f}' = (f_1,\ldots,f_n, 1_{A_1},\ldots,1_{A_m})$ on echalons $T(\bar{E},\bar{A})$, where 1_{A_i} is the identical mapping.

Now let

$$\bar{P}^1 = (P_1^1,\ldots,P_p^1) \in \beta(A_1^1(\bar{E},\bar{A})) \times \ldots \times \beta(A_p^1(\bar{E},\bar{A})) \quad ,$$

$$\bar{P}^{1\prime} = (P_1^{1\prime},\ldots,P_p^{1\prime}) \in \beta(A_1^1(\bar{E}',\bar{A})) \times \ldots \times \beta(A_p^1(\bar{E}',\bar{A})) \quad ,$$

$$\bar{P}_1 = (P_{11},\ldots,P_{1\ell}) \in \beta(A_{11}(\bar{E},\bar{A})) \times \ldots \times \beta(A_{1\ell}(\bar{E},\bar{A})) \quad ,$$

$$\bar{P}_1' = (P_{11}',\ldots,P_{1\ell}') \in \beta(A_{11}(\bar{E}',\bar{A})) \times \ldots \times \beta(A_{1\ell}(\bar{E}',\bar{A})) \quad ,$$

$$\bar{V} = (V_1,\ldots,V_k) \in E_{i_1} \times \ldots \times E_{i_k} \quad ,$$

$$\bar{V}' = (V_1',\ldots,V_k') \in E_{i_1}' \times \ldots \times E_{i_k}' \quad .$$

Definition 2.3. The mapping $\bar{f} = (f_1,\ldots,f_n): \bar{E} \to \bar{E}'$ is called a canonical morphism of the structural set $(\bar{E};\bar{P}^1,\bar{P}_1,\bar{V})$ into the structural set $(\bar{E}';\bar{P}^{1\prime},\bar{P}_1',\bar{V}')$ if and only if for all $i = 1,\ldots,p$, $j = 1,\ldots,\ell$, and $t = 1,\ldots,k$:

$$\bar{f}^{A_i^1}(P_j^{1\prime}) \subset P_i^1 \quad , \quad \bar{f}_{A_{ij}}(P_{ij}) \subset P_{1j}' \quad , \quad f_{i_t}(V_{i_t}) = V_{i_t}' \quad .$$

From the theorem 2.2 if follows

Theorem 2.3. *Any set of structural sets and their canonical morphisms constitute a category. This category will be called the category of structural sets of the species Σ. By abusing denotation we will often denote this category by Σ.*

In categories of structural sets, it is convenient to use objects of some categories of structural sets instead of some sets. Namely, let $U_i = (E_{i1},\ldots,E_{in_i}; \overline{P}^{-1}_i,\overline{P}_{1i},\overline{V}_i)$, $i = 1,\ldots,n$ be objects of the category Σ_i and $(\overline{P}^{-1},\overline{P}_1,\overline{V})$ is a structure on sets $E_1,\ldots,E_n$ and may be on some of E_{ij}. We may construct the category with objects $(U_1,\ldots,U_n, E_1,\ldots,E_n; \overline{P}^{-1},\overline{P}_1,\overline{V})$ and morphisms

$$(\phi_1,\ldots,\phi_n,f_1,\ldots,f_k)\colon (U_1,\ldots,U_n,E_1,\ldots,E_k; \overline{P}^{-1},\overline{P}_1,\overline{V})$$

$$\to (U'_1,\ldots,U'_n,E'_1,\ldots,E'_k; \overline{P}^{-1}{}',\overline{P}'_1,\overline{V}') ,$$

where $\phi_i\colon U_i \to U'_i$ for $i = 1,\ldots,n$ are morphisms in categories Σ_i and $\overline{f} = (f_1,\ldots,f_k)\colon \overline{E} \to \overline{E}'$ is a morphism for the structure $(\overline{P}^{-1},\overline{P}_1,\overline{V})$. Sets $E_{i1},\ldots,E_{in_i}$, $E_1,\ldots,E_k$ may not be all different. Morphisms $\phi_1,\ldots,\phi_n$ are not arbitrary, namely if $\phi_i = (f_{i1},\ldots,f_{in_i})$ where $f_{ij}\colon E_{ij} \to E'_{ij}$, and $\phi_k = (f_{k1},\ldots,f_{kn_k})$, where $f_{k\ell}\colon E_{k\ell} \to E'_{k\ell}$, and $E_{ij} = E_{k\ell}$, $E'_{ij} = E'_{k\ell}$, then for the morphisms ϕ_i and ϕ_k it must be $f_{ij} = f_{k\ell}$.

2.3 Diagram Schemes, Categories, and Functors

For many mathematical structures, morphisms are usually canonical, for example, for algebraic structures, topology structure, and so on. In this section, we show as an example, that canonical morphisms in Category of Categories are functors.

Definition 2.4. Diagram scheme T is a structural set $(\mathrm{ObT},\mathrm{MrT}; \partial)$, where structure ∂ is determined by a covariant echalon $\partial \in \beta(\mathrm{MrT} \times (\mathrm{ObT}\times\mathrm{ObT}))$, and ∂ is a mapping of MrT into $\mathrm{ObT}\times\mathrm{ObT}$. The set MrT is called a set of arrows or morphisms; an element of MrT is called an arrow or a morphism.

The set ObT is called a set of objects, an element of ObT is called an object. The mapping ∂ may be expressed as $\partial = (\partial_0, \partial_1)$, where $\partial_0 = pr_1\partial$, $\partial_1 = pr_2\partial$.

Remark. In other words, a diagram scheme may be defined as an oriented multigraph $(ObT, MrT; \partial_0, \partial_1)$, where ∂_i: MrT $\to$ ObT, but in category theory the notion "diagram" is used.

If s and t are objects, then the set of morphisms (arrows) "between" s and t is denoted by $Hom_T(s,t)$, i.e., $\partial^{-1}(s,t) = Hom_T(s,t)$ or for $s \in ObT$, $t \in ObT$

$$a \in Hom_T(s,t) \iff (\partial_0 a = s,\ \partial_1 a = t)$$

and this relation is denoted by a: s $\to$ t.

As it is follows from n.2.1, a canonical morphism of $T_1 = (ObT_1, MrT_1; \partial^1)$ into $T_2 = (ObT_2, MrT_2; \partial^2)$ is a pair (d_{Ob}, d_{Mr}) of mappings d_{Ob}: $ObT_1 \to ObT_2$, d_{Mr}: $MrT_1 \to MrT_2$ so that $d_{Mr} \times (d_{Ob} \times d_{Ob})(\hat{\partial}^1) \subset \hat{\partial}^2$, where $\hat{\partial}^i$ for i = 1,2 is the graph of the mapping ∂^i. It is not difficult to show that the equivalent relation for (d_{Ob}, d_{Mr}) to be a canonical morphism is for the diagram

$$\begin{array}{ccc} MrT_1 & \xrightarrow{\ d_{Mr}\ } & MrT_2 \\ \partial^1 \big\downarrow & & \big\downarrow \partial^2 \\ ObT_1 \times ObT_1 & \xrightarrow{\ d_{Ob} \times d_{Ob}\ } & ObT_2 \times ObT_2 \end{array}$$

to be commutative. This relation is equivalent for the diagrams

$$\begin{array}{ccc} MrT_1 & \xrightarrow{d_{Mr}} & MrT_2 \\ \partial_0^1 \downarrow & & \downarrow \partial_0^2 \\ ObT_1 & \xrightarrow{d_{Ob}} & ObT_2 \end{array} \quad , \quad \begin{array}{ccc} MrT_1 & \xrightarrow{d_{Mr}} & MrT_2 \\ \partial_1^1 \downarrow & & \downarrow \partial_1^2 \\ ObT_1 & \xrightarrow{d_{Ob}} & ObT_2 \end{array}$$

to be commutative.

Definition 2.5. A diagram scheme $(ObD, MrD; \partial)$ and mappings

$$Id_D: ObD \to MrD \quad ,$$

$$m^D_{d_1,d_2,d_3}: Hom_D(d_1,d_2) \times Hom_D(d_2,d_3) \to Hom_D(d_1,d_3)$$

for every three objects d_1, d_2, d_3 are called a category, if the conditions (1), (2), and (3) below are fulfilled. Morphism $m^D_{d_1,d_2,d_3}(U,V)$, where $U: d_1 \to d_2$, $V: d_2 \to d_3$ is called a composition of the morphisms U, V and is denoted by $V \circ U$; the mapping $m^D_{d_1,d_2,d_3}$ is called a law of composition in the category D.

The conditions to be fulfilled are:

1. $\partial_0(Id_D(d)) = \partial_1(Id_D(d)) = d$.

2. The composition law is associative, i.e., if $U: d_1 \to d_2$, $V: d_2 \to d_3$, and $W: d_3 \to d_4$, then $W \circ (V \circ U) = (W \circ V) \circ U$.

3. For any morphism $U: d \to d_1$

$$U \circ Id_D(d) = Id_D(d_1) \circ U = U \quad .$$

We see that category D is a structural set $(\mathrm{ObD},\mathrm{MrD};\ \partial,\mathrm{Id}_D,m^D)$, where structures ∂, Id_D, m^D are determined by covariant echalons

$$\partial \in \beta(\mathrm{MrD} \times (\mathrm{ObD} \times \mathrm{ObD})) \quad ,$$

$$\mathrm{Id}_D \in \beta(\mathrm{ObD} \times \mathrm{MrD}) \quad ,$$

$$m^D \in \beta((\mathrm{ObD} \times \mathrm{ObD} \times \mathrm{ObD}) \times \beta((\mathrm{MrD} \times \mathrm{MrD}) \times \mathrm{MrD})) \quad ,$$

∂ is a mapping of MrD into $\mathrm{ObD} \times \mathrm{ObD}$, Id_D is a mapping of ObD into MrD, m^D is a mapping of $\mathrm{ObD} \times \mathrm{ObD} \times \mathrm{ObD}$ into $\beta((\mathrm{MrD} \times \mathrm{MrD}) \times \mathrm{MrD})$. The value $m^D(d_1,d_2,d_3)$ (it is often denoted by m^D) is a mapping (or a functional graph) of $\mathrm{Hom}_D(d_1,d_2) \times \mathrm{Hom}_D(d_2,d_3)$ into $\mathrm{Hom}_D(d_1,d_3)$ and the conditions (1), (2), and (3) must be fulfilled. A canonical morphism of category $D_1 = (\mathrm{ObD}_1,\mathrm{MrD}_1;\ \partial^1,\mathrm{Id}_{D_1},m^{D_1})$ into category $D_2 = (\mathrm{ObD}_2,\mathrm{MrD}_2;\ \partial^2,\mathrm{Id}_{D_2},m^{D_2})$ is a pair (d_{Ob},d_{Mr}) of mappings $d_{Ob}\colon \mathrm{ObT}_1 \to \mathrm{ObT}_2$, $d_{Mr}\colon \mathrm{MrT}_1 \to \mathrm{MrT}_2$, so that diagrams

$$\begin{array}{ccc}
\mathrm{MrD}_1 & \xrightarrow{\ d_{Mr}\ } & \mathrm{MrD}_2 \\
\partial^1 \downarrow & & \downarrow \partial^2 \\
\mathrm{ObD}_1 \times \mathrm{ObD}_1 & \xrightarrow{\ d_{Ob} \times d_{Ob}\ } & \mathrm{ObD}_2 \times \mathrm{ObD}_2
\end{array} \quad ,$$

$$\begin{array}{ccc}
\mathrm{ObD}_1 & \xrightarrow{\ d_{Ob}\ } & \mathrm{ObD}_2 \\
\mathrm{Id}_{D_1} \downarrow & & \downarrow \mathrm{Id}_{D_2} \\
\mathrm{MrD}_1 & \xrightarrow{\ d_{Mr}\ } & \mathrm{MrD}_2
\end{array} \quad ,$$

$$\begin{array}{ccc}
ObD_1 \times ObD_1 \times ObD_1 & \xrightarrow{d_{Ob} \times d_{Ob} \times d_{Ob}} & ObD_2 \times ObD_2 \times ObD_2 \\
{\scriptstyle m^{D_1}} \downarrow & & \downarrow {\scriptstyle m^{D_2}} \\
\beta((MrD_1 \times MrD_1) \times MrD_1) & \xrightarrow{((d_{Mr} \times d_{Mr}) \times d_{Mr})^{\beta}} & \beta((MrD_2 \times MrD_2) \times MrD_2)
\end{array}$$

are commutative. The last condition is equivalent for diagram

$$\begin{array}{ccc}
Hom_{D_1}(d_1,d_2) \times Hom_{D_1}(d_2,d_3) & \xrightarrow{d_{Mr} \times d_{Mr}} & Hom_{D_2}(d_{Ob}(d_1),d_{OB}(d_2)) \times Hom_{D_2}(d_{Ob}(d_2),d_{Ob}(d_3)) \\
{\scriptstyle m^{D_1}} \downarrow & & \downarrow {\scriptstyle m^{D_2}} \\
Hom_{D_1}(d_1,d_3) & \xrightarrow{d_{Mr}} & Hom_{D_2}(d_{Ob}(d_1),d_{Ob}(d_3))
\end{array}$$

to be commutative; in this diagram, the mapping d_{Mr} is a restriction on $Hom_D(d_i, d_j)$ for suitable i and j.

Commutativeness of the second and third diagrams means that

1. for every $d \in ObD_1$

$$Id_{D_2}(d_{Ob}(d)) = d_{Mr}(Id_{D_1}(d)) \quad ;$$

2. for every three objects d_1, d_2, d_3 of ObD and every two morphisms U: $d_1 \rightarrow d_2$, V: $d_2 \rightarrow d_3$

$$d_{Mr}(V \circ U) = d_{Mr}(V) \circ d_{Mr}(U) \quad .$$

The last relations are the standard definition of the functor.

2.4 Category of Hospitals

We constructed the hospital structure in point 1.1. By using the technique described in 2.2, we introduce now morphisms for hospital structure and, therefore, the category of hospitals.

In the structure H defined in (1.11) we have the following primitive sets: R, B, Σ, DS, and D. We introduce therefore the following mappings: f_R: R $\rightarrow$ R', f_B: B $\rightarrow$ B', f_Σ: $\Sigma \rightarrow \Sigma'$, f_{DS}: DS $\rightarrow$ DS', f_D: D $\rightarrow$ D'. The array $f_H = (f_R, f_B, f_\Sigma, f_{DS}, f_D)$ is a canonical morphism from a hospital H (as an element of the hospital system) into a hospital H' if

$$f_Z \times \beta(f_Y)(X_Y^Z) \subset X_{Y'}^{Z'}$$

for $X_Y^Z = P_B^R$, or F_{DS}^B, or F_{DS}^Σ, or F_{DS}^D, or $F_B^{P(D,DS)}$. In these conditions, we denoted $\beta(f) = f_\beta$, and for Z = P(D,DS) we should use $f_D \times f_{DS}$. These conditions are equivalent to the following:

the diagrams

$$\begin{array}{ccc} Z & \xrightarrow{f_Z} & Z' \\ {\scriptstyle X_Y^Z}\downarrow & & \downarrow{\scriptstyle X_{Y'}^{Z'}} \\ \beta(Y) & \xrightarrow{\beta(Y)} & \beta(Y') \end{array}$$

are commutative.

For hospitals as systems, we assume that the element base described by the graph with one element, one input, and one output is the same for all hospitals. Therefore, element bases of all hospitals in our case are the same to within the labels. Consider first the patient structure $P^S = (P,DS;\ T_{DS}^P)$ and assigning array structure $A^P = (f_B^P, f_{DS}^P, f_\Sigma^P, f_D^P)$ holding the conditions (1.12) - (1.18). The primitive data in patient structure are sets P and DS. Therefore, if $f_P\colon P \to P'$, then a canonical morphisms from a patient structure $P^S = (P,DS;\ T_{PS}^P)$ into $P^{S'} = (P',DS';\ T_{DS'}^{P'})$ is a pair (f_P, f_{DS}) such that

$$f_P \times \beta(f_{DS})(T_{DS}^P) \subset T_{DS'}^{P'}$$

or, which is the same, the diagram

$$\begin{array}{ccc} P & \xrightarrow{f_P} & P' \\ {\scriptstyle T_{DS}^P}\downarrow & & \downarrow{\scriptstyle T_{DS'}^{P'}} \\ \beta(DS) & \xrightarrow{\beta(f_{DS})} & \beta(DS') \end{array}$$

is commutative.

The primitive data in assigning array A^P are sets P, B, DS, Σ, and D, and therefore a canonical morphism from structure A^P into $A^{P'}$ is an array $(f_P, f_B, f_{DS}, f_\Sigma, f_D)$ such that

$$f_P \times f_B(f_B^P) \subset f_{B'}^{P'}\ , \qquad f_P \times \beta(f_Y)(f_Y^P) \subset f_{B'}^{P'} \quad \text{for } Y = DS, \Sigma, D$$

or, which is the same, the diagrams

$$\begin{array}{ccc} P & \xrightarrow{f_P} & P' \\ {\scriptstyle f_B^P}\downarrow & & \downarrow{\scriptstyle f_{B'}^{P'}} \\ B & \xrightarrow{f_B} & B' \end{array} \quad , \qquad \begin{array}{ccc} P & \xrightarrow{f_P} & P' \\ {\scriptstyle f_Y^P}\downarrow & & \downarrow{\scriptstyle f_{Y'}^{P'}} \\ \beta(Y) & \xrightarrow{\beta(f_Y)} & \beta(Y') \end{array}$$

are commutative.

Now, we can introduce canonical morphisms from an information base $IB_{SYS}^H = (\Gamma, S^H)$ for a hospital H into an information base $IB_{SYS}^{H'} = (\Gamma', S^{H'})$ for a hospital H'. The element base Γ' is the same as Γ to within the labels. Primitive data for the information base $IB_{SYS}^H = (\Gamma,\ S^H)$ are sets R, B, Σ, DS, D, and for every available state $s \in S^H$ three sets P_I, P_H, and P_O. We denote this state s as $s(P_I, P_H, P_O)$. We denote also

$$f_{P_I} : P_I \to P'_I\ , \quad f_{P_H} : P_H \to P'_H\ , \quad f_{P_O} : P_O \to P'_O \quad .$$

A canonical morphism from a structure IB_{SYS}^H into $IB_{SYS}^{H'}$ is an array of mappings $(f_R, f_B, f_\Sigma, f_{DS}, f_D)$ such that it is a morphism from the hospital H into the hospital H'; and for every $s(P_I, P_H, P_O) \in S^H$ there exits a state $s'(P'_I, P'_H, P'_O) \in S^{H'}$ so that:

1. the pair (f_{P_I}, f_{DS}) is a morphism from the patient structure $pr_1 s(P_I, P_H, P_O) = (P_I, DS; T_{DS}^{P_I})$ into the patient structure $pr_1 s'(P'_I, P'_H, P'_O) = (P'_I, DS'; T_{DS'}^{P'_I})$;
2. the pair (f_{P_H}, f_{DS}) is a morphism from the patient structure $pr_1 pr_2 s(P_I, P_H, P_O) = (P_H, DS; T_{DS}^{P_H})$ into the patient structure $pr_1 pr_2 s'(P'_I, P'_H, P'_O) = (P'_H, DS'; T_{DS}^{P'_H})$; the array $(f_{P_H}, f_B, f_{DS}, f_\Sigma, f_D)$ is a morphism from the assigning array $A^{P_H} = (f_B^{P_H}, f_{DS}^{P_H}, f_\Sigma^{P_H}, f_D^{P_H})$ into the assigning array $A^{P'_H} = (f_{B'}^{P'_H}, f_{DS}^{P'_H}, f_\Sigma^{P'_H}, f_{D'}^{P'_H})$;
3. the pair (f_{P_O}, f_{DS}) is a morphism from the patient structure $pr_1 pr_3 s(P_I, P_H, P_O) = (P_O, DS; T_{DS}^{P_O})$ into the patient structure $pr_1 pr_3 s'(P'_I, P'_H, P'_O) = (P'_O, DS'; T_{DS'}^{P'_O})$; the array $(f_{P_O}, f_B, f_{DS}, f_\Sigma, f_D)$ is a morphism from the assigning array A^{P_O} into the assigning array $A^{P'_O}$.

3. UNIVERSAL QUERY LANGUAGE ON FINITE PRIMITVE SETS

In the previous section, we defined structural sets of a given type. The general construction contains transportable relations as additional axioms. In this section we are going to demonstrate techniques for construction of the whole "world" of objects and relations on the basis of a given set of primitive sets. Relations constructed can be used as transportable relations. We will assume that primitive sets are finite sets. The "world" of objects and relations will be called the Universal Query Language on Finite Primitive Sets. We will demonstrate the techniques developed only for the case of one primitive set. We define the notions "theorem" and "existence quantifier" and show that the set of relations is a logical and quantifier theories. We introduce, further, the relation "belong" (ε), the primitive relation of the Set Theory, and show that the set of relations holds some properties of the set theory.

3.1 Languages with Weights. Significant and Balanced Words.

In this and the next point we develop techniques for construction objects, sets of objects, and relations of our "world." In this point we follow N. Bourbaki, Elements of Mathematics, "Theory of Sets," Addison-Wesley Publishing Co., 1968.

Let S be a nonempty set. Elements of S will be called *signs* and the set S will be called an *alphabet*. The set of all finite words in the alphabet S is denoted by S*, that is, $S^* = \cup_{n\geq 0} S^n$, where S^0 is an empty word ϕ. The number of signs in the word $A \in S^*$ is denoted by $\ell(A)$ and is called the *length* of A. We have the following binary operation on S*:

$$\cdot : S^* \times S^* \to S^*: (A,B) \to A \cdot B \quad ,$$

where the word $A \cdot B$ which will be denoted as AB, is the concatenation of A and B and is the word that consists of A followed by the word B.

We have

$$\ell(AB) = \ell(A) + \ell(B) \quad .$$

A *weight on* S is a function

$$n: S \to \mathbf{N}: s \to n(s) \quad ,$$

$n(s)$ is called the weight of the sign s. We extend the function n on the set S* as follows

$$n: S^* \to \mathbf{N}: (s_1, s_2, \ldots, s_k) \to \sum_{j=0}^{k} n(s_k).$$

We have

$$n(AB) = n(A) + n(B) \quad .$$

The structure (S; n) will be called a language with a weight.

Definition 3.1. Given a structure (S; n). A sequence (A_j) $1 \leq j \leq k$ of words from S* is called a significant sequence if for every word A_j one of the following is true:

i) A_j is a sign of the zero weight;

ii) there are p words $A_{i_1}, A_{i_2}, \ldots, A_{i_p}$ from the sequence with $i_e < j$ for $e = 1,2,\ldots,p$, and a sign f of the weight p such that

$$A_j = fA_{i_1}A_{i_2}\ldots A_{i_p}$$

Words of significant sequences are called significant words.

We have immediately

Proposition 3.1. *If* $A_1, A_2, \ldots, A_p$ *are significant words and* f *is a sign of the weight* p, *then the word* $fA_1A_2\ldots A_p$ *is significant.*

Definition 3.2. Given a structure (S; n). A word $A \in S^*$ is called a balanced word if

i) $\ell(A) = n(A) + 1$

ii) for every proper initial segment B of the word A (that is, A = BC and $C \neq \phi$)

$\ell(B) \leq n(B)$.

We have

Proposition 3.2. *Every balanced word* A *can be represented as*

$$A = fA_1A_2\ldots A_p$$

where all A_i *and balanced words and* $n(f) = p$.

Proposition 3.3. *Given a structure* (S; n). *A word* $A \in S^*$ *is significant if and only if it is balanced.*

Corollary 3.1. *Every signifianct word* A *can be represented as* $A = fA_1A_2\ldots A_p$ *where all* A_i *are significant and* $n(f) = p$. *This representation is unique.*

3.2 Complex-Schemes. Complexes

Definition 3.3. Given n+1 intervals $[1,m_i]$, $i = 1,2,\ldots,n+1$ of integer numbers (that is, $[1,m_i] \subset \mathbf{N}$ for $i = 1,\ldots,n+1$), where $m_1 = 1$, that is, $[1,m_1] = [1,1] = \{1\}$, and given n functions $(f_i)_{i\in[1,n]}$, where for every $i = 1,2,\ldots,n$

$$f_i\colon [1,m_{i+1}] \to [1,m_i]$$

such that

$$f_i(m) < f_i(k) \Rightarrow m < k \quad \text{for} \quad m,k \in [1,m_{i+1}] \quad .$$

The sum of the family $([1,m_i])_{i\in[1,n+1]}$ and the family of functions $(f_i)_{i\in[1,n]}$, that is, the object

$$\left(\coprod_{i\in 1}^{n+1} [1,m_i],\ (f_i)_{i\in[1,n]}\right)$$

is called a complex-scheme.

Proposition 3.4. *The graph*

$$G = \left\{((f_i(m),i),(m,i+1)) : m \in [1,m_{i+1}],\ i = 1,2,\dots,n\right\}$$

on the set $\coprod_{i=1}^{n+1} [1,m_i]$ *is a tree-graph with layers*

$$\{1\}\times\{1\},\quad [1,m_2]\times\{2\},\quad [1,m_3]\times\{3\},\dots,[1,m_{n+1}]\times\{n+1\}.$$

Example 3.1.

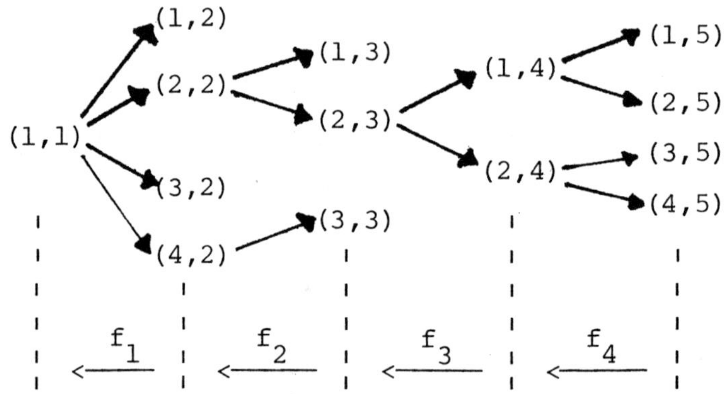

f_1: $\{(1,2),(2,2),(3,2),(4,2)\} \to \{(1,1)\}$;

f_2: $\{(1,3),(2,3)\} \to \{(2,2)\}$, $\{(3,3)\} \to \{(4,2)\}$;

f_3: $\{(1,4),(2,4)\} \to \{(2,3)\}$;

f_4: $\{(1,5),(2,5)\} \to \{(1,4)\}$, $\{(3,5),(4,5)\} \to \{(2,4)\}$.

Definition 3.4. Let $\Gamma = (\coprod_{i=1}^{n+1} [1,m_i], (f_i)_{i\in[1,n]})$ be a complex-scheme. A function

$$f: \coprod_{i=1}^{n+1} [1,m_i] \to X$$

will be called a complex on X and often will be denoted by X_f^Γ.

If we would like to depict a structure X_f^Γ on X we will depict elements of X on vertices of corresponding graph for Γ.

Example 3.2. Let X = {a,b,c,d,e}, Γ be a complex-scheme from example 3.1, the complex f is

x	(1,1)	(1,2)	(2,2)	(3,2)	(4,2)	(1,3)	(2,3)
f(x)	a	a	b	c	c	d	e

x	(3,3)	(1,4)	(2,4)	(1,5)	(2,5)	(3,5)	(4,5)
f(x)	d	b	e	b	a	a	c

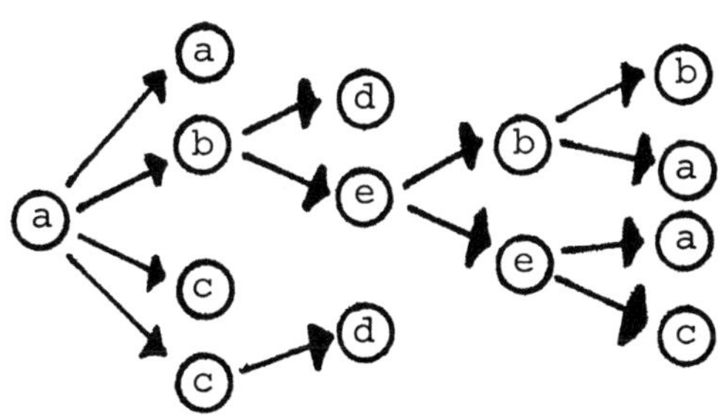

Definition 3.5. Let Γ be a complex-scheme. We say that the element (m_1,n_1) is less than the element (m_2,n_2) and denote

$$(m_1,n_1) <_\Gamma (m_2,n_2)$$

if $n_1 = n_2$ and $m_1 < m_2$, or

if $n_1 \neq n_2$ but

i) there is a path from (m_1,n_1) to (m_2,n_2) (in this case, $n_1 < n_2$), or

ii) there is no path from (m_1,n_1) to (m_2,n_2) but there are vertices (k_1,p) and (k_2,p) such that $k_1 < k_2$, and there are paths from (k_1,p) to (m_1,n_1) and from (k_2,p) to (m_2,n_2).

Theorem 3.1. *The relation* $(m_1,n_1) \leq_\Gamma (m_2,n_2)$, *that is, the relation* $(m_1,n_1) <_\Gamma (m_2,n_2)$ or $(m_1,n_1) = (m_2,n_2)$ *(that is,* $m_1 = m_2$, $n_1 = n_2$) *is a total order on the set* $\coprod_{i=1}^{n+1} [1,m_i]$. *The element* (1.1) *is a minimal element.*

Example 3.3. For the complex-scheme from example 3.1, we have

$$(1,1) < (1,2) < (2,2) < (1,3) < (2,3) < (1,4)$$
$$< (1,5) < (2,5) < (2,4) < (3,5) < (4,5)$$
$$< (3,2) < (4,2) < (3,3) \quad .$$

If X_f^Γ is a complex on S, then we will often write the element f(x) for $x \in \coprod_{i=1}^{n+1} [1,m_i]$ in the order $\leq_\Gamma$ by assigning level (that is $pr_2 x$) to the element f(x).

Example 3.4. The complex X_f^Γ from the example 3.2, may be represented as

1a,2a,2b,3d,3e,4b,5b,5a,4e,5a,5c,2c,2c,3d .

3.3 Significant Words and Complexes

For a graph, an arc beginning at a vertex will be called *an output of the vertex*, and an arc ending at a vertex will be called an *input of the vertex*. *Indegree of a vertex* is the number of inputs of this vertex, and *outdegree of a vertex* is the number of outputs of this vertex.

Definition 3.6. Let (S; n) be a language with a weight n and $\Gamma = (\coprod_{i=1}^{k+1} [1,m_i], (f_i)_{i\in[1,k]})$ be a complex-scheme. A complex S_f^Γ will be called a complex on (S; n) if for every vertex $(j,\ell) \in \coprod_{i=1}^{k+1} [1,m_i]$ the outdegree of this vertex equals the weight of the sign assigned to this vertex, that is,

$$\text{outdegree}(j,\ell) = n(f(j,\ell)).$$

We now assign a complex-scheme and a complex to every significant word of the language (S; n). Given p complex-schemes $\Gamma_1,\Gamma_2,\ldots,\Gamma_p$, that is, for j = 1,2,...,p we have sets $\coprod_{i=1}^{n_j+1} [1,m_{i,j}]$ and functions $f_{ij}: [1,m_{i+1,j}] \to [1,m_{i,j}]$ for $i = 1,2,\ldots,n_j$.

By $\otimes(\Gamma_1,\Gamma_2,\ldots,\Gamma_p)$ we denote the complex-scheme with vertices

$$\left\{\{(1,1)\} \cup \coprod_{i=2}^{\max\limits_{1\le j\le p} n_j+2} [1, \sum_{j=1}^{p} m_{i-1,j}]\right\}$$

and the family of functions

$$(f_i)_{i\in[1,\max\limits_{1\le j\le p} n_j+1]}$$

$$f_1: [1,p] \to \{1\}, \quad f_{i+1}: [1, \sum_{j=1}^{p} m_{i+1,j}] \to [1, \sum_{j=1}^{p} m_{i,j}],$$

$$i = 1,2,\ldots,\max_{1\le j\le p} n_j$$

where

$$f_i\Big|_{[m_{i+1,j}+1,m_{i+1,j+1}]} : [m_{i+1,j}+1,m_{i+1,j+1}] \rightarrow [m_{i,j}+1,m_{i,j+1}]$$

and

$$f_i\Big|_{[m_{i+1,j}+1,m_{i+1,j+1}]}(r) = f_{i,j+1}(r - m_{i+1,j}) \quad .$$

Complex-schemes $\Gamma_1,\Gamma_2,\ldots,\Gamma_p$ in $\otimes(\Gamma_1,\Gamma_2,\ldots,\Gamma_p)$ will be called *complex-subschemes*.

Example 3.5.

Γ_1

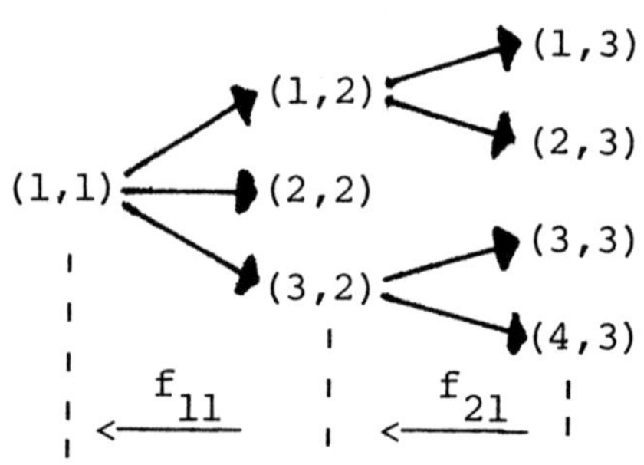

Γ_2

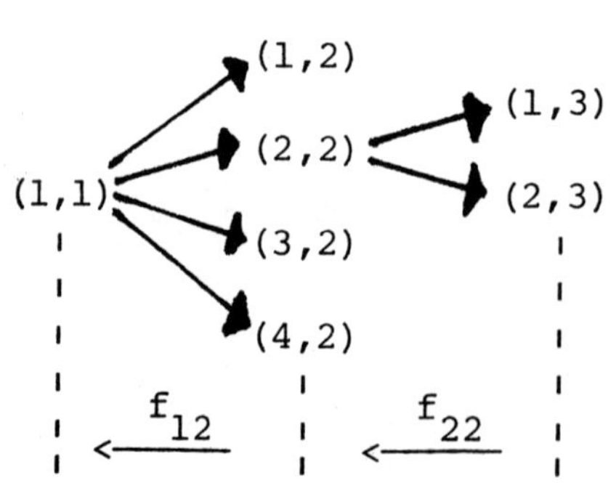

$\otimes(\Gamma_1,\Gamma_2)$

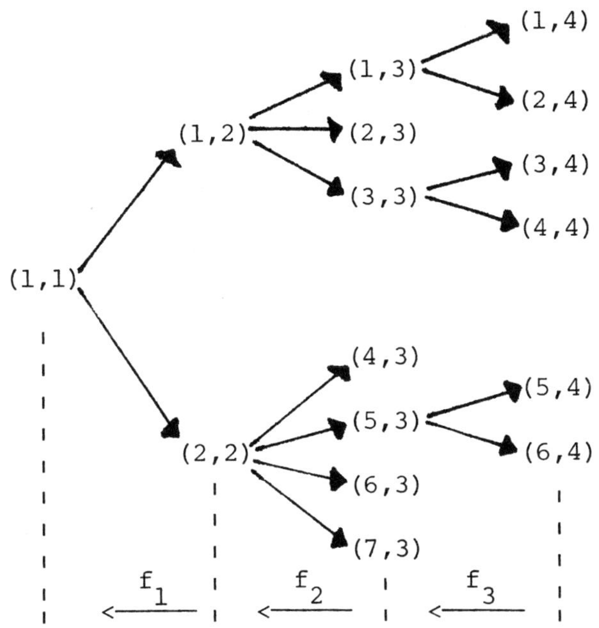

Let (S; n) be a language with the weight n. For every sign s with the zero weight, that is, n(s) = 0, we assign the complex-scheme {(1,1)} and the complex 1s, that is,

$$f\colon \{(1,1)\} \to f((1,1)) = s \quad .$$

Let us assume that we assigned complexes S_f^{Γ} for all significant words containing less than n signs and let A be a significant word of the form

$$A = fA_1A_2\ldots A_p \quad (n(f) = p)$$

for which $A_1,A_2,\ldots,A_p$ contain less than m signs. Therefore, we have complexes $S_{f_1}^{\Gamma_1},S_{f_2}^{\Gamma_2},\ldots,S_{f_p}^{\Gamma_p}$. We have the complex-scheme $\otimes(\Gamma_1,\Gamma_2,\ldots,\Gamma_p)$; the outdegree of the vertex (1,1) of which equals p.

We construct the function

$$f: \mathrm{Ver}(\otimes(\Gamma_1, \Gamma_2, \ldots, \Gamma_p)) \to S$$

where $\mathrm{Ver}(\otimes(\Gamma_1, \Gamma_2, \ldots, \Gamma_p))$ is the set of vertices of the complex-scheme $\otimes(\Gamma_1, \ldots, \Gamma_p)$, and

$$f((1,1)) = f\ , \quad f\Big|_{\mathrm{Ver}(\Gamma_1)} = f_1\ , \quad f\Big|_{\mathrm{Ver}(\Gamma_2)} = f_2, \ldots,$$

$$f\Big|_{\mathrm{Ver}(\Gamma_p)} = f_p \quad ,$$

$\mathrm{Ver}(\Gamma_j)$ is the set of vertices of the complexe-scheme Γ_j. The function f is a complex on (S; n) which will be denoted by $\otimes(f, S_{f_1}^{\Gamma_1}, S_{f_2}^{\Gamma_2}, \ldots, S_{f_p}^{\Gamma_p})$. This complex is assigned to the significant word $A = fA_1A_2 \ldots A_p$.

Example 3.6. Let $S = \{+,*,0,1,2,3,4,5,6,7,8,9\}$, $n(+) = n(*) = 2$, $n(i) = 0$ for $i = 1,2,\ldots,9$. Given two significant words +1*0+14 and +*+2567 with the following corresponding complexes

$S_{f_1}^{\Gamma_1}$:

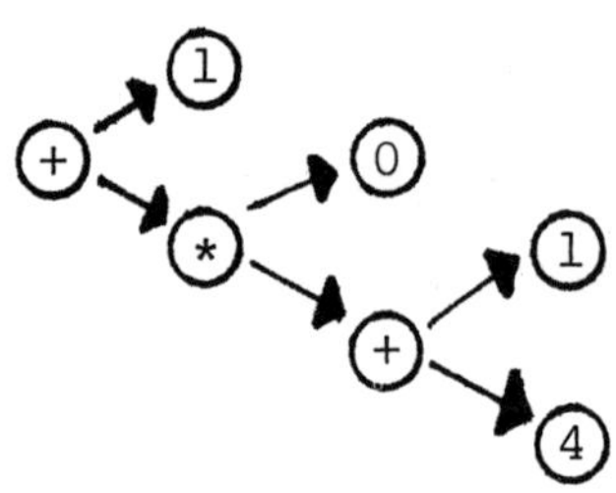

$S_{f_2}^{\Gamma_2}$:

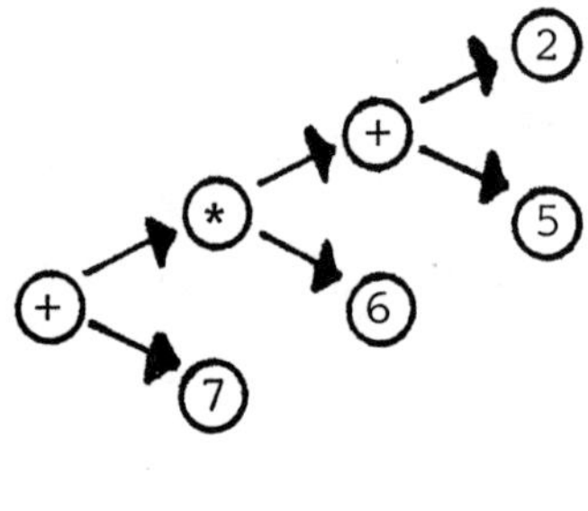

The significant word *+1*0+14+*+2567 corresponds to the following complex

$\otimes(*, S_{f_1}^{\Gamma_1}, S_{f_2}^{\Gamma_2})$:

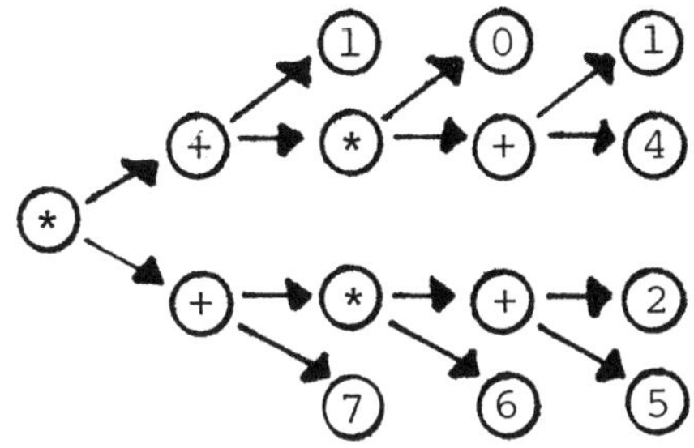

Proposition 3.5. *Let* S_f^{Γ} *be a complex on* (S; n) *and* A_Γ *be the set of leaves of* Γ. *Then*

$$a \in A_\Gamma \iff (a \in \mathrm{Ver}(\Gamma) \text{ and } n(a) = 0) \quad .$$

That is, signs of S *assigned to leaves and only those signs have zero weight.*

If S is a finite set and $s_1, s_2, \ldots, s_m$ are signs from S of non-zero weight and $a_1, a_2, \ldots, a_p$ are signs from S of zero weight, then we will denote often the language (S; n) as

$$L(s_1, s_2, \ldots, s_m;\ a_1, a_2, \ldots, a_p) \quad .$$

We will consider often languages $L(s_1, \ldots, s_m;\ a_1, \ldots, a_p)$ signs $s_1, \ldots, s_m$ of which have a different weight, that is $n(s_i) = n(s_j)$ follows $s_i = s_j$. Every significant word of such a language is determined uniquely by a graph Γ and by a function $f\colon A_\Gamma \to \{a_1, a_2, \ldots, a_p\}$ from the set A_Γ of leaves of Γ into the set $\{a_1, a_2, \ldots, a_p\}$ of signs of zero weight. We will denote a significant word corresponding to a graph Γ and a function f by Γ(f). Corresponding notation will be used also for infinite set S where it will be appropriate.

3.4 Cartesian Complexes on a Given Set of Primitive Data and Variables. Constants and Terms

Let E be a set. We will call often its elements as primitive data. Consider the language

$$L(\mathfrak{I}^i,\ i \geq 2;\ E)$$

where for the weight function

$$n:\ E \cup \{\mathfrak{I}^i:\ i \geq 2\} \to \mathbb{N}$$

we have $n(x) = 0$ for $x \in E$ and $n(\mathfrak{I}^i) = i$.

If $y_1, y_2, \ldots, y_i$ are significant words, then we denote

$$\mathfrak{I}^i y_1 y_2 \ldots y_i = (y_1, y_2, \ldots, y_i) \quad .$$

Significant words of this language will be called cartesian complexes on the set E. Every cartesian complex is represented by a tree-graph Γ which vertices have, if any, not less that two outputs and by a function f from leaves of Γ into the set E, it is denoted by $\Gamma(f)$. The set of all significant words for a given tree-graph Γ, that is, the set of all functions f from the set of leaves of Γ into E, will be denoted by $\Gamma(E)$. This set is described only by the tree-graph Γ. If for a significant word $\mathfrak{I}^i y_1 y_2 \ldots y_i = (y_1, y_2, \ldots, y_i)$ $\Gamma_1, \Gamma_2, \ldots, \Gamma_i$ are tree-graphs corresponding to significant words $y_1, y_2, \ldots, y_i$, then the set of all significant words of this type with fixed tree-graphs $\Gamma_1, \ldots, \Gamma_i$ is denoted by

$$\Gamma_1(E) \times \Gamma_2(E) \times \ldots \times \Gamma_i(E) \quad .$$

If we introduce symbols X^i for $i \geq 2$ and correspondence $\Gamma_1(E) \times \Gamma_2(E) \times \ldots \times \Gamma_i(E) = X^i \Gamma_1(E) \Gamma_2(E) \ldots \Gamma_i(E)$ for all sets of significant words with a given tree-graph, then all these sets are significant words in the language

$$L(X^i, i \geq 2; \{E\})$$

with the weight $n(X^i) = i$, $n(E) = 0$.

We extend now the language $L(\mathfrak{I}^i, i \geq 2; E)$ by introducing the set of variables V_E, that is the set of symbols different from symbols from E, $E \cap V_E = \phi$. The weight of $v \in V_E$ is zero. Elements of V_E will be called variables on E.

We have therefore the language

$$L(\mathfrak{I}^i, i \geq 2; E \cup V_E) = T(E, V_E) \quad .$$

Every element from $T(E,V_E)$ is called a term. Therefore, a term is determined by a tree Γ every vertex of which has if any not less than two outputs and by a function f from the set A_Γ of leaves of Γ into $E \cup V_E$; this term will be denoted by $\Gamma(f)$.

A term $\Gamma(f) \in T(E,V_E)$ is called a *constant* if $f(A_\Gamma) \subset E$, that is, if the corresponding significant word does not contain variables. A non-constant term will be called a *(variable) term*.

Let $A_\Gamma = A_\Gamma^c \cup A_\Gamma^v$, where $f(A_\Gamma^c) \subset E$ and $f(A_\Gamma^v) \subset V_E$. It is obvious that $A_\Gamma^c \cap A_\Gamma^v = \phi$. A function $\phi\colon f(A_\Gamma^v) \to E$ is called an *assignment (of variables from* $f(A_\Gamma^v)$*)*. The constant $\Gamma(\overline{f})$, where

$$\overline{f}\colon A_\Gamma \to E \ ; \quad \overline{f}\Big|_{A_\Gamma^c} = f \ ; \quad \overline{f}\Big|_{A_\Gamma^v} = \phi \circ f$$

is called the *value of the term* $\Gamma(f)$ *at assignment* ϕ.

3.5 Equivalence Relations

We consider now the language $L(=; T(E,V_E))$, where the weight of the symbol $=$ equals 2 and the weight of every element from $T(E,V_E)$ equals 0.

We say that a term t is *of the type* Γ if it is determined as $\Gamma(f)$ for some f. We say that terms t_1 and t_2 are *of the same type* if they are determined as $\Gamma(f_1)$ and $\Gamma(f_2)$ for some Γ, f_1, and f_2. We introduce now the language $Eq(T(E,V_E))$ as follows:

i) every element of $Eq(T(E,V_E))$ is an element of $L(=;\ T(E,V_E))$, that is, $Eq(T(E,V_E)) \subset L(=;\ T(E,V_E))$, and has the form $= t_1t_2$ for some $t_1,t_2 \in T(E,V_E)$. We will denote this term as $t_1 = t_2$ and call it an equivalence relation;

ii) $t_1 = t_2 \in Eq(T(E,V_E)) \iff$ (t_1 and t_2 are terms of the same type).

Defintion 3.7. Let t_1 and t_2 be terms of the same type determined by $\Gamma(f_1)$ and $\Gamma(f_2)$. If t_1 and t_2 are constants, then the equivalence relation $t_1 = t_2$ is called a theorem if and only if $f_1 = f_2$, that is the terms t_1 and t_2 are the same. If t_1 and t_2 are variable terms, then the equivalence relation $t_1 = t_2$ is called a theorem if and only if for every assignment ϕ

$$\phi\colon f_1(A^v_{\Gamma,1}) \cup f_2(A^v_{\Gamma,2}) \to E$$

where $A_\Gamma = A^c_{\Gamma,1} \cup A^v_{\Gamma,1} = A^c_{\Gamma,2} \cup A^v_{\Gamma,2}$, the equivalence relation $\bar{t}_1 = \bar{t}_2$ for the values of t_1 and t_2 at assignment ϕ is a theorem. The values $\bar{t}_1$ and $\bar{t}_2$ for the assignment ϕ are determined correspondingly as $\Gamma(\bar{f}_1)$ and $\Gamma(\bar{f}_2)$, where for $i = 1,2$

$$\bar{f}_i\Big|_{A^c_{\Gamma,i}} = f_i\Big|_{A^c_{\Gamma,i}}, \quad \bar{f}_i\Big|_{A^v_{\Gamma,i}} = \phi \circ f_i\Big|_{A^v_{\Gamma,i}}.$$

Theorem 3.2. *If the set* E *contains more than one element, then an equivalence relation* $t_1 = t_2$ *is a theorem if and only if the terms* t_1 *and* t_2 *are the same, that is, they are determined by the same* $\Gamma(f)$.

Corollary 3.2. *For every element* e *from* E *and for every variable* x *from* V_E, *the equivalence relations*

$$e = e\ , \quad x = x$$

are theorems. If E *contains more than two elements, then for every two different elements* e_1 *and* e_2 *from* E *and for two different variables* x_1 *and* x_2 *from* V_E, *the equivalence relations*

$$e_1 = e_2\ , \quad x_1 = x_2\ , \quad x_1 = e_1$$

are not theorems. If $t_1 = t_2$ *is a theorem, then* $t_2 = t_1$ *is a theorem. If* $t_1 = t_2$ *and* $t_2 = t_3$ *are theorems, then* $t_1 = t_3$ *is a theorem.*

3.6 Relations. Theorems. Classes of Relations.

We consider now the language $L(\neg, \vee;\ Eq(T(E,V_E)))$, where the weight of the symbol $\neg$ (NOT) equals 1, the weight of the symbol $\vee$ (OR) equals 2, and for every equivalence relation $t_1 = t_2$ the weight of the relation equals zero.

We denote

$R_1 \vee R_2$ instead of $\vee R_1R_2$ (R_1 or R_2)

$R_1 \Rightarrow R_2$ instead of $\vee\neg R_1R_2$ (R_1 implies R_2)

$R_1 \wedge R_2$ instead of $\neg((\neg R_1)\vee(\neg R_2))$ (R_1 and R_2)

$R_1 \Leftrightarrow R_1$ instead of $(R_1 \Rightarrow R_2) \wedge (R_2 \Rightarrow R_1)$
(R_1 equivalent R_2)

An element of the language $L(\neg, \vee;\ Eq(T(E,V_E)))$ will be called a relation. If R is a relation, then it is determined by a tree-graph Γ each element of which has one or two outputs and by a function $\phi\colon A_\Gamma \to Eq(T(E,V_E))$, where A_Γ is the set of leaves of Γ. Therefore, for every $a \in A_\Gamma$, the value $\phi(a)$ is an equivalence relation. If $\phi(a)$ is an equivalence relation for two constants for every a from A_Γ, then the relation R will be called a *sentence*. A relation which is not a sentence will be called a (*variable*) *relation*.

Let R be a variable relation, Γ be the corresponding tree-graph for the relation, A_Γ be the set of leaves for Γ, and $\phi\colon A_\Gamma \to Eq(T(E,V_E))$ be the corresponding function. We denote

$\Gamma_{\phi(a)}$ is the tree describing an equivalence relation $\phi(a)$

$A_{\phi(a)}$ is the set of leaves of $\Gamma_{\phi(a)}$

$f_{\phi(a),i}\colon A_{\phi(a)} \to E \cup V_E$

for i = 1,2 are functions which determine the terms of equivalence relation $\phi(a)$. We also have

$$A_{\phi(a)} = A^c_{\phi(a),i} \cup A^v_{\phi(a),i} \qquad (i = 1,2)$$

where

$$f_{\phi(a),i}(A^c_{\phi(a),i}) \subset E\ , \qquad f_{\phi(a),i}(A^v_{\phi(a),i}) \subset V_E \quad .$$

The function

$$\Phi\colon \bigcup_{a\in A_\Gamma} \bigcup_{i\in[1,2]} f_{\phi(a),i}(A^v_{\phi(a),i}) \to E$$

is called, as above, *an assignment*. The set $\bigcup_{a\in A_\Gamma} \bigcup_{i\in[1,2]} f_{\phi(a),i}(A^v_{\phi(a),i})$ is a set of variables in the relation R.

The sentence $\overline{R}$ determined by Γ and the function

$$\overline{\phi}: A_\Gamma \to Eq(T(E,V_E)) \quad ,$$

where for the equivalence relation $\overline{\phi}(a)$ we have

$$\Gamma_{\overline{\phi}(a)} = \Gamma_{\phi(a)}$$

$$f_{\overline{\phi}(a),i}\Big|_{A^c_{\phi(a)}} = f_{\phi(a),i}\Big|_{A^c_{\phi(a)}} \quad ,$$

$$f_{\overline{\phi}(a),i}\Big|_{A^v_{\phi(a),i}} = \Phi \circ f_{\phi(a),i}\Big|_{A^v_{\phi(a),i}} \qquad \text{for } i = 1,2$$

will be called the *value of* R *at assignment* Φ.

Definition 3.8. A sentence $\neg R$ is called a theorem if and only if the sentence R is not a theorem; a sentence $R_1 \vee R_2$ is called a theorem if and only if the sentence R_1 or the sentence R_2 is a theorem. A (variable) relation $\neg R$ is called a theorem if and only if the value of the relation $\neg R$ at any assignment Φ is a theorem. A (variable) relation $R_1 \vee R_2$ is called a theorem if and only if the value of the relation $R_1 \vee R_2$ at any assignment Φ is a theorem.

Definition 3.8 should be considered an inductive definition.

Theorem 3.8. *The set of all theorems of the language* $L(\neg, \vee; Eq(T(E,V_E)))$ *is a logical theory, that is, if* P, R, *and* Q *are relations, then the relations*

i) $(P \vee Q) \Rightarrow P$

ii) $P \Rightarrow (P \vee R)$

iii) $(P \vee R) \Rightarrow (R \vee P)$

iv) $(P \Rightarrow R) \Rightarrow ((Q \vee P) \Rightarrow (Q \vee R))$

are theorems.

Theorem 3.4. *A sentence* R *is a theorem if and only if the sentence* $\neg$R *is not a theorem. If a variable relation* R *is a theorem, then the relation* $\neg$R *is not a theorem, but not vice versa. The relation* R$\vee$($\neg$R) *is a theorem for every* R.

Definition 3.9. We will say that two relations R_1 and R_2 are equivalent if and only if the relation $R_1 \Longleftrightarrow R_2$ is a theorem. If R_1 and R_2 are equivalent, then we denote $R_1 \equiv R_2$. The relation $R_1 \equiv R_2$ is an equivalence relation on the language $L(\neg, \vee; Eq(T(E,V_E)))$. If R is a relation, then the class of all equivalent relations will be denoted $\|R\|$.

Theorem 3.5. *The set of classes of equivalent relations is a boolean algebra with minimal element* $\|R \wedge (\neg R)\|$ *and maximal element* $\|R \vee (\neg R)\|$.

3.7 Substitution. Quantifiers.

Let t be a term. Therefore, it is determined by a tree Γ and a function $f\colon A_\Gamma \to E \cup V_E$. Let x be a variable from V_E and τ be an element from E or V_R. Consider the function

$$(\tau|x)f\colon A_\Gamma \to E \cup V_E \quad ,$$

so that

$$(\tau|x)f\Big|_{A_\Gamma - f^{-1}(x)} = f\Big|_{A_\Gamma - f^{-1}(x)} , \qquad (\tau|x)f\Big|_{f^{-1}(x)} = \tau \quad .$$

This function defines the term which is the term t where every entrance of the variable x is substituted by the element τ. This new term will be denoted by $(\tau|x)t$. Let, now, R be a relation. Therefore, it is determined by a tree Γ and a function $\phi\colon A_\Gamma \to Eq(T(E,V_E))$. The value $\phi(a)$ for

$a \in A_\Gamma$ is an equivalence relation $t_1 = t_2$. Consider the function

$$(\tau|x)\phi\colon A_\Gamma \to Eq(T(E,V_E))$$

where $(\tau|x)\phi(a)$ is the equivalence relation $(\tau|x)t_1 = (\tau|x)t_2$ for every $a \in A_\Gamma$. This relation will be denoted by $(\tau|x)R$. We assume now that the set E is a finite set and $E = \{e_1, e_2, \ldots, e_k\}$.

Definition 3.10. The relation

$$\bigvee_x R \overset{\text{def}}{=} (e_1|x)R \vee (e_2|x)R \vee \ldots \vee (e_k|x)R$$

is called "there exists x from E such that R"; the relation

$$\bigwedge_x R \overset{\text{def}}{=} \neg\bigvee_x(\neg R)$$

is called "for every x from E R." The symbols $\bigvee_x$ and $\bigwedge_x$ are called the existence and the universal quantifiers.

In definition 3.10 the relations $R_1 \vee R_2 \vee \ldots \vee R_k$ represents the relation $(\ldots(R_1 \vee R_2) \vee R_3) \vee \ldots) \vee R_k)$.

Theorem 3.6. *If* $\tau \in E \cup V_E$ *and* $x \in V_E$, *then the relation*

$$(\tau|x)R \Rightarrow \bigvee_x R$$

is a theorem.

3.8 The Relation "Belong"

Let t be a term determined by a graph Γ and a function $f\colon A_\Gamma \to E \cup V_E$. We denoted this term also by $\Gamma(f)$, and the set of all terms with fixed Γ was denoted by $\Gamma(E)$. We assume now that the set E is a finite set and denote the set of all assignments $\phi: A_\Gamma^V \to E$ as $\{\phi_1, \phi_2, \ldots, \phi_m\}$ and the corresponding values for t as $\bar{t}_1, \bar{t}_2, \ldots, \bar{t}_m$ or as $\Gamma(\bar{f}_1), \Gamma(\bar{f}_2), \ldots, \Gamma(\bar{f}_m)$ where

$$\bar{f}_i\Big|_{A_\Gamma^C} = f\Big|_{A_\Gamma^C}, \quad \bar{f}_i\Big|_{A_\Gamma^C} = \phi_i \circ f\Big|_{A_\Gamma^C}.$$

Definition 3.11. The relation

$$\Gamma(f) \in \Gamma(E) \overset{\text{def}}{=} ((\Gamma(f) = \Gamma(\bar{f}_1)) \vee (\Gamma(f) = \Gamma(\bar{f}_2)) \vee \ldots \vee (\Gamma(f) = \Gamma(\bar{f}_m))),$$

or which is the same,

$$t \in \Gamma(E) \overset{\text{def}}{=} (t = \bar{t}_1 \vee t = \bar{t}_2 \vee \ldots \vee t = \bar{t}_m)$$

is called "the term t (or $\Gamma(f)$) belongs to (or is an element of) the set $\Gamma(E)$."

Theorem 3.7. *The relation* $\Gamma(f) \in \Gamma(E)$ *is a theorem.*

Corollary 3.3. *For every element* e *from the set* E *and every variable* x *from* V_E, *the relations*

$$e \in E, \quad x \in E$$

are theorems. For every set of variables $x_1, x_2, \ldots, x_n$ *from* V_E, *the relation*

$$(x_1, x_2, \ldots, x_n) \in E \times E \times \ldots \times E \overset{\text{def}}{=} E^n$$

is a theorem.

Different constructions equivalent to those from the set theory may be developed by using the relation "belong"; and many equivalent theorems describing properties of these constructions may be proved. Consideration of these constructions and properties is not a goal of this paper.

4. SYSTEMS AND THEIR INFORMATION BASES

4.1 Category of Systems

We view a system to be a collection of connected simultaneously acting elements. So, a system consists of a connected set of *elements, element base* B, *a space of available states of element base* S^B, *a space of time scales* Θ, *and a space of processes* Π, each of which takes place on its own time scale. Hence, a system η is the collection of four objects

$$\eta = (B, S^B, \Theta, \Pi) \quad .$$

Objects B, S^B, Θ, Π are really structural sets. The element base B consists of a set A of elements, a set I of inputs, a set O of outputs, and a set C of connections for any two elements $a \in A$ and $b \in A$. Therefore, the element base is a structural set

$$B = (A, I, O, C; \partial_0, \partial_1) \quad ,$$

where $\partial_0 \colon O \cup C \to A$, $\partial_1 \colon I \cup C \to A$ are mappings determining elements $\partial_0(c)$ and $\partial_1(c)$ that are connected by the connection $c \in C$. The definition of element base is an extension of the definition of a diagram scheme: $B = (\mathrm{Ob}B, \mathrm{Mr}B; \partial_0, \partial_1)$ where $\mathrm{Ob}B = A$, $\mathrm{Mr}B = I \cup O \cup C$ and ∂_0, ∂_1 are *partial* mappings of MrB into ObB. A canonical morphism of the elements base $B_1 = (A_1, I_1, O_1, C_1; \partial_0^1, \partial_1^1)$ into the elements base $B_2 = (A_2, I_2, O_2, C_2; \partial_0^2, \partial_2^2)$ consists of four mappings

$\phi_B = (f_A, f_I, f_O, f_C)$; $f_A\colon A_1 \to A_2$, $f_I\colon I_1 \to I_2$, $f_O\colon O_1 \to O_2$, $f_C\colon C_1 \to C_2$ such that diagrams

$$\begin{array}{ccc} I_1 & \xrightarrow{f_I} & I_2 \\ \partial_1^1 \downarrow & & \downarrow \partial_1^2 \\ A_1 & \xrightarrow{f_A} & A_2 \end{array}, \quad \begin{array}{ccc} O_1 & \xrightarrow{f_O} & O_2 \\ \partial_0^1 \downarrow & & \downarrow \partial_0^2 \\ A_1 & \xrightarrow{f_A} & A_2 \end{array}, \quad \begin{array}{ccc} C_1 & \xrightarrow{f_C} & C_2 \\ \partial_i^1 \downarrow & & \downarrow \partial_i^2 \\ A_1 & \xrightarrow{f_A} & A_2 \end{array},$$

$i = 1,2$

are cummutative. In this diagram the mappings $\partial_{0,1}^{1,2}$ are corresponding restrictions. Objects of the category of element bases will be often denoted by $B(A,I,O,C)$. The definition of the nature of an object x will be stated by assigning to it the space of its states E_x. So, to the set X of elements we may assign the mapping $E_X\colon X \to \beta(\Sigma_X)$: $x \to E_X(x) = E_x$, where E_x is a space of states of x. We call Σ_X as a space of states of the set X. A state s of X as a whole is a mapping $s\colon X \to \Sigma_X$, so that $s(x) \in E_x$ for every $x \in X$. So, the set of available states of elements of X is a set S of such mappings

$$s\colon X \to \Sigma_X\colon x \to s(x) \quad (s(x) \in E_x) \quad .$$

Hence, the *space of available states of elements of* X is a structural set

$$S^X = (X, \Sigma_X;\ E_X, S_X) \quad ,$$

where $E_X \in \beta(X \times \beta(\Sigma_X))$, $S_X \in \beta(\beta(X \times \Sigma_X))$; E_X is a mapping of X into $\beta(\Sigma_X)$ and every element $s \in S_X$ is a mapping of X into Σ_X, so that $s(x) \in E_X(x) = E_x$.

A canonical morphism of the space $S^{X_1} = (X_1, \Sigma_{X_1}; E_{X_1}, S_{X_1})$ into the space $S^{X_2} = (X_2, \Sigma_{X_2}; E_{X_1}, S_{X_2})$ is a pair $\phi_X = (f_X, f_{\Sigma_X})$: $S^{X_1} \to S^{X_2}$ of mappings $f_X\colon X_1 \to X_2$, $f_{\Sigma_X} : \Sigma_{X_1} \to \Sigma_{X_2}$ so that the diagrams

$$\begin{array}{ccc} X_1 & \xrightarrow{f_X} & X_2 \\ {\scriptstyle E_{X_1}}\downarrow & & \downarrow{\scriptstyle E_{X_2}} \\ \beta(\Sigma_{X_1}) & \xrightarrow{\beta(f_{\Sigma_X})} & \beta(\Sigma_{X_2}) \end{array} \qquad \text{for every } s_1 \in S_{X_1} \text{ there exists } s_2 \in S_{X_2} \qquad \begin{array}{ccc} X_1 & \xrightarrow{f_X} & X_2 \\ {\scriptstyle s_1}\downarrow & & \downarrow{\scriptstyle s_2} \\ \Sigma_{X_1} & \xrightarrow{f_{\Sigma_X}} & \Sigma_{X_2} \end{array}$$

are commutative. For element base $\hat{B} = (A, I, O, C; \partial_0, \partial_1)$ by denoting $B = A \coprod I \coprod O \coprod C$ (the sum of the sets A, I, O, and C), we have the space of available states $S^B = (B, \Sigma_B; E_B, S_B)$. We will often denote a space of available states by $S(B, S_B)$.

The space of time scales Θ consists of a set T of time scales, a set of times E_τ on every scale $\tau \in T$, and a total order G_τ on the set of times E_τ. Therefore, *the space of time scales* is a structural set

$$\Theta = (T, \Sigma_T; E_T, G) \quad ,$$

where $E_T \in \beta(T \times \Sigma_T)$ is a mapping of T into $\beta(\Sigma_T)$, $E_T(\tau) = E_\tau$ for every $\tau \in T$; $G \in \beta(T \times \beta(\Sigma_T \times \Sigma_T))$ is a mapping of T into $\beta(\Sigma_T \times \Sigma_T)$ and $G(\tau) = G_\tau$ is a total order on E_τ.

Canonical morphism $\phi_\Theta\colon \Theta_1 \to \Theta_2$ of the space of time scales $\Theta_1 = (T_1, \Sigma_{T_1}; E_{T_1}, G_1)$ into $\Theta_2 = (T_2, \Sigma_{T_2}; E_{T_2}, G_2)$ is a pair of mappings $\phi_\Theta = (f_T, f_{\Sigma_T})$, $f_T\colon T_1 \to T_2$, $f_{\Sigma_T} : \Sigma_{T_1} \to \Sigma_{T_2}$ for which the diagrams

$$\begin{array}{ccc} T_1 & \xrightarrow{f_T} & T_2 \\ {\scriptstyle E_{T_1}}\downarrow & & \downarrow{\scriptstyle E_{T_2}} \\ \beta(\Sigma_{T_1}) & \xrightarrow{\beta(f_{\Sigma_T})} & \beta(\Sigma_{T_2}) \end{array} ,$$

$$\begin{array}{ccc} T_1 & \xrightarrow{\qquad f_T \qquad} & T_2 \\ {\scriptstyle G_1}\downarrow & & \downarrow{\scriptstyle G_2} \\ \beta(\Sigma_{T_1} \times \Sigma_{T_2}) & \xrightarrow{\beta(f_{\Sigma_T} \times f_{\Sigma_T})} & \beta(\Sigma_{T_2} \times \Sigma_{T_2}) \end{array}$$

are commutative. The structural set of spaces of time scales will often be denoted by $\Theta(T,\Sigma_T)$.

The additional reasonable axiom may be suggested for the space of time scales: yesterday, present, and future are the same for the system, i.e., there is no such sequence $(t_i,t_{i+1})_{i\in[0,n]}$ so that $t_i \in E_\tau$, $t_{i+1} \in E_\tau$, for some τ, $t_i \leq t_{i+1}$ and $t_0 \geq t_{n+1}$. In this case, there exists the ordered set Σ and $E_\tau \subset \Sigma$ for all $\tau \in T$. In other words, Σ_T is a set of totally ordered subsets of the ordered set Σ. A process π in the set S with the set of times E_τ is a mapping π: $E_\tau \to S$. And so, if $\Theta = (T,\Sigma_T;\ E_T,G)$ is a space of time scales, then the set $\hat{\Pi}$ of processes in the set S is an element $\hat{\Pi} \in \beta(\beta(\Sigma_T \times S))$, where for every $\pi \in \hat{\Pi}$ there is unique $\tau \in T$ and π: $E_\tau \to S$.

Hence, the *space of processes in the set* S is a structural set

$$\Pi = ((T,\Sigma_T;\ E_T,G),S;\ \hat{\Pi}) = (\Theta,S;\ \hat{\Pi}) \quad .$$

Canonical morphism $\phi_\Pi\colon \Pi_1 \to \Pi_2$ of the space of processes $\Pi_1 = (\Theta_1,S_1;\ \hat{\Pi}_1)$ into the space of processes $\Pi_2 = (\Theta_2,S_2;\ \hat{\Pi}_2)$ consists of three mappings $\phi_\Pi = (f_T,f_{\Sigma_T},f_S)$, where $\phi_\Theta = (f_T,f_{\Sigma_T})\colon \Theta_1 \to \Theta_2$ is a morphism in the category of the spaces of time scales, and for every $\pi \in \hat{\Pi}_1$ there exists a process $\pi_2 \in \hat{\Pi}_2$ so that the diagram

$$\begin{array}{ccc} T_1 & \xrightarrow{f_T} & T_2 \\ {\scriptstyle E_{T_1}}\downarrow & & \downarrow{\scriptstyle E_{T_2}} \\ \beta(\Sigma_{T_1}) & \xrightarrow{\beta(f_{\Sigma_T})} & \beta(\Sigma_{T_2}) \end{array} \qquad\qquad \begin{array}{ccc} \Sigma_{T_1} & \xrightarrow{f_{\Sigma_T}} & \Sigma_{T_2} \\ {\scriptstyle \pi_1}\downarrow & & \downarrow{\scriptstyle \pi_2} \\ S_1 & \xrightarrow{f_S} & S_2 \end{array}$$

is commutative. A structural set of the space of processes in S will often be denoted by $\Pi(\Theta(T,\Sigma_T),S)$ or $\Pi(\Theta,S)$. Hence the system consists of the following four objects

$$\begin{aligned} \eta &= (B,S^B,\Theta,\Pi) \\ &= (B(A,I,O,C),S(B,S_B),\Theta(T,\Sigma_T),\Pi(\Theta(T,\Sigma_T),S^B)) \quad . \end{aligned}$$

Therefore, a canonical morphism $\phi_\eta\colon \eta_1 \to \eta_2$ of the system $\eta_1 = (B_1,S^{B_1},\ \Theta_1,\Pi_1)$ into the system $\eta_2 = (B_2,S^{B_2},\ \Theta_2,\Pi_2)$ is a set of seven mappings $\phi_\eta = (f_A,f_I,f_O,f_C,f_{\Sigma_B},f_T,f_{\Sigma_T})$ so that $\phi_B = (f_A,f_I,f_O,f_C)\colon B_1 \to B_2$ is a canonical morphism of B_1 into B_2; $\phi_S = (f_B,f_{\Sigma_B})\colon S^{B_1} \to S^{B_2}$ $(f_B\colon A_1 \cup I_1 \cup O_1 \cup C_1 \to A_2 \cup I_2 \cup O_2 \cup C_2$, $f_B(x) = f_A(x)$ or $f_I(x)$ or $f_O(x)$ or $f_C(x)$

for corresponding $x \in A_1$ or $x \in I_1$ or $x \in O_1$ or $x \in C_1$) is a corresponding morphism of S^{B_1} into S^{B_1}; $\phi_\Theta = (f_T, f_{\Sigma_T})$: $\Theta_1 \to \Theta_2$ is a canonical morphism of Θ_1 into Θ_2; $\phi_\Pi = (f_T, f_{\Sigma_T}, (f_B \times f_{\Sigma_B})^\beta)$: $\Pi_1 \to \Pi_2$ is a canonical morphism of Π_1 into Π_2.

4.2 Examples

In this section, we consider, as examples, such systems as automata, Turing machines, and difference schemes containing von Neuman cellular automata.

4.2.1 Automata

The element base and spaces of states are defined on the graph below:

Here, X is a space of states of the input, Y is a space of states of the output, and Q is a space of states of the automaton as an element. The space of available states of this system is $S = X \times Q \times Y$. The space of time scales consists of all intervals $\tau = [n_1, n_2]$ in the set of integers $\mathbf{Z}$, so that $n_1 \leq n_2$. The space of processes consists of all mappings π: $[n_1, n_2] \to X \times Q \times Y$: $n \to (x_n, q_n, y_n)$ defined for any given initial state $s_0 = (x_{n_1}, q_{n_1}, y_{n_1})$ and process on the input $(x_{n_1}, x_{n_1+1}, x_{n_1+2}, \ldots, x_{n_2})$ so that $(x_{n+1}, q_{n+1}, y_{n+1}) = (x_{n+1}, \lambda(x_{n+1}, q_n), \delta(x_{n+1}, q_n))$. Here λ: $X \times Q \to Q$, δ: $X \times Q \to Y$. The automaton, therefore, is a deterministic system for every given input process. An automaton A is usually defined as $A = (X,Q,Y,\lambda,\delta)$ (in the structure sense $A = (X,Q,Y;\ \lambda,\delta))$, and morphisms of the automaton $A_1 = (X_1,Q_1,Y_1;\ \lambda_1,\delta_1)$ into the automaton $A_2 = (X_2,Q_2,Y_2;\ \lambda_2,\delta_2)$ is an array of mappings $f = (f_1,f_2,f_3)$ so that $f_1\colon X_1 \to X_2$, $f_2\colon Q_1 \to Q_2$, $f_3\colon Y_1 \to Y_2$ and the diagram

$$\begin{array}{ccccc}
Q_1 & \xleftarrow{\ \delta_1\ } & X_1 \times Q_1 & \xrightarrow{\ \lambda_1\ } & Y_1 \\
{\scriptstyle f_2}\downarrow & & \downarrow{\scriptstyle f_1 \times f_2} & & \downarrow{\scriptstyle f_3} \\
Q_2 & \xleftarrow{\ \delta_1\ } & X_2 \times Q_2 & \xrightarrow{\ \lambda_1\ } & Y_2
\end{array}$$

is commutative. Of course, these morphisms are morphisms in the category of automaton systems but our category has additional morphisms.

4.2.2 Turing Machines

The element base of a machine consists of the reading head (or head) H and a tape infinite in both directions; each element, or cell, of a tape is defined by an integer $z \in \mathbf{Z}$. So, the element base is a set $\{H\} \times \mathbf{Z}$ (there are no arcs). The space of states of H is a set $Q = \{q_0,\ldots,q_m\}$, the space of state of any $z \in \mathbf{Z}$ is a set $A = \{a_0,\ldots,a_n\}$. The space of available states is defined as follows. All cells but a finite number are in the state a_0, the head may be in any state $q_i \in Q$ and observes the cell z. So, any available state of the machine may be uniquely described by the word

$$x_{z-\ell}x_{z-\ell+1}x_{z-\ell+2}\cdots x_{z-1}y_z x_z x_{z+1}\cdots x_{z+k} \quad , \qquad (*)$$

where $x_i \in A$, $y_z \in Q$, $x_{z-\ell} \neq a_0$, $x_{z+k} \neq a_0$. This word means that the head observes the cell z and is in the state y_z; the state of any cell $t < z-\ell$ and $t > z+k$ is a_0; $x_{z-\ell}$ is the first and x_{z+k} is the last state not equal to a_0. So, the space of available state S is equivalent to all words of the (*) type. We will call two words $x_{z_1-\ell_1}\dots x_{z_1-1}y_{z_1}x_{z_1}\dots x_{z_1+k_1}$ and $x_{z_2-\ell}\dots x_{z_2-1}y_{z_2}x_{z_2}\dots x_{z_2+k_2}$, and corresponding states $s_1 \in S$ and $s_2 \in S$ equivalent if $\ell = \ell_1 = \ell_2$, $k_1 = k_2 = k$ and $x_{z_1-i} = x_{z_2-i}$ for $i = 1,\dots,\ell$, $y_{z_1} = y_{z_2}$, $x_{z_1+i} = x_{z_2+i}$ for $i = 1,\dots,k$. The spaces of time scales T and processes Π for the system are defined by the following. Relation R "s_1 and s_2 are equivalent" is equivalence. Let $\hat{S} = S/R$ be a factor set. Any time scale τ and processes π are defined for a given class $\hat{s} \in \hat{S}$ and the matrix T. The machine matrix T consists of elements $t(a_i,q_j)$ for elements $a_0,\dots,a_n$, $q_1,\dots,q_m$;

$$t(a_i,q_j) = q_r(a_i,q_j)U(a_i,q_j) \quad ,$$

where $U(a_i,q_j)$ is equal to $a_p(a_i,q_j)$, i.e., to some "letter" a_p for p depending on a_i and q_j, or to "r" (it means "right") or to "l" (it means "left").

Matrix T is an operator of $\hat{S}$ into $\hat{S}$:

$$T(x_{z-\ell}\dots x_{z-1}q_j a_i x_{z+1}\dots x_{z+k})$$

$$= \begin{cases} x_{z-\ell}\dots x_{z-1}q_r a_p x_{z+1}\dots x_{z+k} & \text{if } U(a_i,q_j) = a_p \\ x_{z-\ell}\dots x_{z-1}a_i q_r x_{z+1}\dots x_{z+k} & \text{if } U(a_i,q_j) = r \\ x_{z-\ell}\dots q_r x_{z-1}a_i x_{z+k}\dots x_{z+k} & \text{if } U(a_i,q_j) = \ell \end{cases}$$

Any process $\pi \in \Pi$ is defined for a given initial state $s^0 = x^0_{z-\ell}\ldots x^0_{z-1}y^0_z x^0_z\ldots x^0_{z+k}$ and consists of sequences s^0, Ts^0, $T^2s_0, \ldots, T^n s_0$ such that the state of the head in $T^n s_0$ equals q_0, and for any $i < n$ the state of the head in $T^i s_0$ is not equal to q_0. So, q_0 is the "stop" state.

Therefore, for any $\hat{s} \in \hat{S} = S/R$ we have the positive number $n(\hat{s})$ (may be equal to the infinity), so that s, Ts, $T^2 s, \ldots, T^{n(\hat{s})} s$, where $s \in \hat{s}$, is a process of the system. Hence, the space of time scales of the system consists of all integer intervals $[n, n(\hat{s})+n]$ for any $n \in \mathbf{Z}$ and $\hat{s} \in S$. The space of processes Π consists of all mappings $\pi \in \Pi$

$$\pi\colon [n, n(\hat{s})+n] \to S\colon k \to \pi(k) = T^{k-n}s \quad ,$$

where $s \in \hat{s}$. A Turing machine, therefore, is a deterministic system. For example, if $A = \{0,1\}$, $Q = \{q_0, q_1, q_2\}$

$$T = \begin{array}{c|cc} & q_1 & q_2 \\ \hline 0 & q_2 r & q_0 1 \\ 1 & q_1 r & q_2 r \end{array}$$

and $\hat{s} = q_1 11$, then

$$T(\ldots 0q_1 110\ldots) = (\ldots 01q_1 10\ldots) \quad ,$$

$$T^2(\ldots 0q_1 110\ldots) = (\ldots 011q_1 0\ldots) \quad ,$$

$$T^3(\ldots 0q_1 110\ldots) = (\ldots 0110q_2 00\ldots) \quad ,$$

$$T^4(\ldots 0q_1 110\ldots) = (\ldots 0110q_0 10\ldots) \quad ,$$

and so the sequence

$$(\ldots 0q_1 1100\ldots),(\ldots 01q_1 100),(\ldots 11q_1 00\ldots),$$
$$(\ldots 0110q_2 10\ldots),(\ldots 0110q_0 10\ldots)$$

is a process of the system.

4.2.3 Difference Schemes

Difference schemes are a large class of systems very important in mathematics and physics. For example, von Neuman cellular automata are a subclass of difference schemes; finite difference equations are another important subclass. For simplicity, I define difference schemes on a plane.

The element base of a system is a set of all integer pairs (m,n), i.e., $B = \mathbf{Z} \times \mathbf{Z} = \mathbf{Z}^2$. The space of available states for any element, for simplicity, is supposed to be the same and equal to a set E, i.e., $E_{(m,n)} = E$. The space of available states is

$$S = \prod_{(m,n)\in \mathbf{Z}^2} E_{(m,n)} = E^{\mathbf{Z} \times \mathbf{Z}}$$

The space of time scales consists of all intervals in $\mathbf{Z}$, i.e., all subsets $[n,m] \subset \mathbf{Z}$, $n \in \mathbf{Z}$, $m \in \mathbf{Z}$, $m \geq n$. For definition of the space of processes, I define the notion "template" and vicinity of an element (m,n). A template V is a finite subset in $\mathbf{Z}^2$, i.e., $V \subset \mathbf{Z}^2$ containing the origin (the element (0,0)). The vicinity of an element (m,n) is a subset

$$V_{(m,n)} = (m,n) + V = \{(i,k) : i = m+t,\ k = n+r,\ (t,r) \in V\} \quad .$$

Given a mapping $F: E^V \to E$. By using the bijection

$K_{(m,n)}: V \to V_{(m,n)}: (i,j) \to (i+m, j+n)$, we define the function $F_{(m,n)}: E^{V_{(m,n)}} \to E: u \to F_{(m,n)}(u) = F(u \circ K_{(m,n)})$. If $s \in S$, i.e., $s: \mathbf{Z}^2 \to E$ and $s|_{V_{(m,n)}}$ is the restriction of s on $V_{(m,n)}$, then for every (m,n) we have $F_{(m,n)}(s|_{V_{(m,n)}}) = F(s|_{V_{(m,n)}} \circ K_{(m,n)}) \in E$. Therefore, if $\bar{s}: \mathbf{Z}^2 \to E$: $(m,n) \to F(s|_{V_{(m,n)}} \circ K_{(m,n)})$, then $\bar{s} \in S$, so, we can define the operator,

$$T: S \to S: s \to T(s) \quad ,$$

where $T(s)(m,n) = F(s|_{V_{(m,n)}} \circ K_{(m,n)})$.

Any process π of the system is defined for any initial state $s^0 \in S$ by the relation

$$\pi: [n_1,n_2] \to S: n \to T^{n-n_1}(s_0) \quad .$$

A difference scheme, therefore, is a deterministic system. Let, for example,

$$\frac{[U_{\tau+\theta}(x,y) - U_{\theta}(x,y)]}{\theta}$$

$$= \frac{[U_\tau(x+h_1,y) - 2U_\tau(x,y) + U_\tau(x-h_1,y)]}{h_1^2}$$

$$+ \frac{[U_\tau(x,y+h_2) - 2U_\tau(x,y) + U_\tau(x,y-h_2)]}{h_2^2}$$

for $x = mh_1$, $y = nh_2$, $\tau = t\theta$, $(m,n) \in \mathbf{Z}^2$, $t \in [n_1,n_2]$, is a difference approximation for differential equation $\partial U/\partial\tau = \partial^2 U/\partial x^2 + \partial^2 U/\partial y^2$ $(\tau_0 \leq \tau \leq \tau_1,\ (x,y) \in \mathbf{R}^2)$.

Defining the template $V = \{(0,0),(1,0),(0,1),(-1,0),(0,-1)\}$, a mapping $\ell: V \to \mathbf{R}$, i.e., $\ell \in \mathbf{R}^V$ and $\ell = (\ell_{(0,0)},\ell_{(1,0)},\ell_{(0,1)},\ell_{(-1,0)},\ell_{(0,-1)})$, and the mapping $F: \mathbf{R}^V \to \mathbf{R}: \ell \to F(\ell)$, where

$$F(\ell) = \left(1 - 2\Theta\,\frac{h_1^2 + h_2^2}{h_1^2 + h_2^2}\right)\ell_{(0,0)} + \left(\frac{\Theta}{h_1^2}\right)(\ell_{(1,0)} + \ell_{(-1,0)}) + \left(\frac{\Theta}{h_2^2}\right)(\ell_{(0,1)} + \ell_{(0,-1)}) \quad ,$$

we transfer the difference equation to the difference scheme with the processes

$$s_{t+1}(m,n) = \left(1 - 2\Theta\,\frac{h_1^2 + h_2^2}{h_1^2 + h_2^2}\right)s_t(m,n) + \left(\frac{\Theta}{h_1^2}\right)(s_t(m+1,n)+s_t(m-1,n)) + \left(\frac{\Theta}{h_2^2}\right)(s_t(m,n+1) + s_t(m,n-1)) \quad ,$$

where

$$s_t(m,n) = U_{t\Theta}(mh_1,nh_2) \quad .$$

4.3 Information Base of a System

In the point 4.1, we defined a system as the collection of four objects: an element base B, a space of available states of element base S^B, a space of time scales O, and a space of processes Π. The pair

$$IB_{SYS} = (B,S^B)$$

is a particular case of the information base of a system defined in point 1.2, where we assigned a structure to every element of an element base, and a space of available states was a subcategory of a product of a family of categories. We see that a system may be defined by its information base, by a space of time scales, and by a space of processes, that is,

$$\eta = (IB_{SYS}, \Theta, \Pi) \quad .$$

In the two sections which follow, we will consider particular cases of information bases as we defined them in this point and analyze them in detail. The analysis of more general information bases needs further development of the theory of category of categories of structural sets.

5. COMPOSITION OF FUNCTIONAL INFORMATION BASES

In this section, I consider a particular case of information bases called functional information bases. This type of information bases describes many technical systems such as electronic circuits, pipelines with pumps and/or heaters. Elements of such systems are described by functional equations, and the functional transformation of the whole system is of a particular interest. I formulate the composition problem for general functional systems, construct the class of specific systems of functional equations describing the information base of the whole system, make the analysis of the algebraic structure of these systems of equations, and introduce and analyze the specific class of functional transformations, which are used for solutions of these systems of equations. The language of the theory of structural sets is very convenient for formulation of the problem and results.

5.1 Composition of Information Bases. Functional Information Bases.

Let $(B_j)_{j\in J} = (A_j, C_j, I_j, O_j;\ \partial_0^j, \partial_1^j)_{j\in J}$ be a family of element bases and

$$A = \coprod_{j\in J} A_j\,,\quad C_1 = \coprod_{j\in J} C_j\,,\quad \overline{I} = \coprod_{j\in J} I_j\,,\quad \overline{O} = \coprod_{j\in J} O_j\,.$$

Let $C_2 \subset \overline{O}$ and a mappong $f\colon C_2 \to \overline{I}$ be an injection; this function f will be called a *scheme of composition*.

Definition 5.1. The element base $B(f) = (A, C, I, O;\ \partial_0, \partial_1)$ where $I = \overline{I} - f(C_2)$, $C = C_1 \cup C_2$, $O = \overline{O} - C_2$, is called the composition of the family $(B_j)_{j\in J}$ of the scheme of composition f if and only if for every $a \in I$ we have $\partial_1(a) = (pr_1 a, \partial_1^{pr_1 a}(pr_2 a))$; for every $b \in O$ we have $\partial_0(b) = (pr_1 b, \partial_0^{pr_1 b}(pr_2 b))$; for every $c \in C$ we have $\partial_i(c) = \partial_i(pr_1 c, \partial_i^{pr_i c}(pr_2 c))$; for every $c \in C_2$ we have $\partial_0(c) = (pr_1 f(c), \partial_0^{pr_1 f(c)}(pr_2 f(c)))$, $\partial_1(c) = (pr_1 c, \partial_1^{pr_1 c}(pr_2 c))$.

The sets A, C, I, and O are called correspondingly the set of elements, connections, inputs, and outputs of the composition.

Example 5.1.

B_1

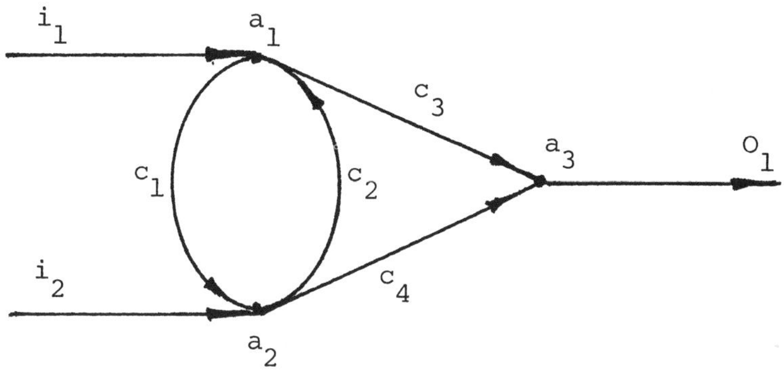

B_2

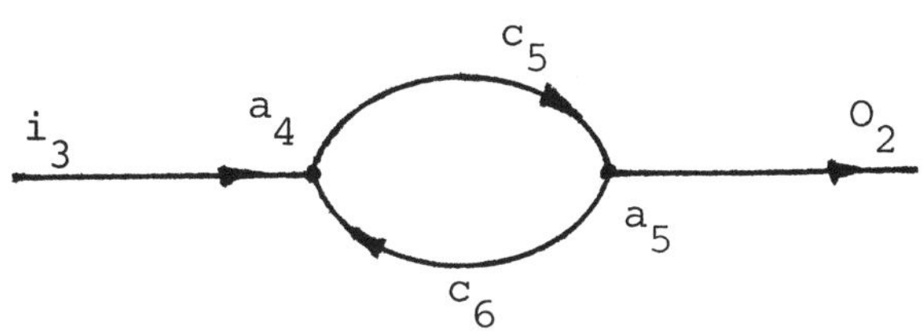

$c_2 = \{O_1\}$

$f(O_1) = i_3$

B(f)

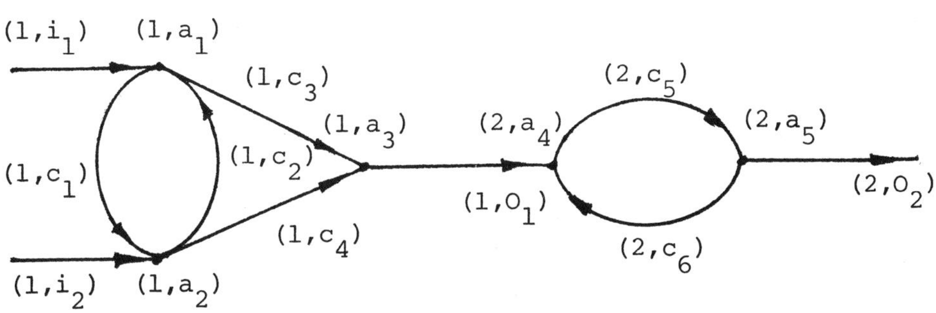

It is obvious that every oriented graph is a composition of its elements.

Let $(B_j, S^{B_j})_{j\in J}$ be a family of information bases. A scheme of composition is called *compatible* with the family of information bases if and only if for every $c \in C_2$ we have $E_{pr_2 c}^{pr_1 c} = E_{pr_2 (c)}^{pr_1 f(c)}$, that is, the spaces of states of the arcs $pr_2 c$ and $pr_2 f(c)$ identified in the composition $B(f)$ are the same.

Definition 5.2. If $(B_j, S^{B_j})_{j\in J}$ is a family of information bases and f is a scheme of composition compatible with this family, then the information base $(B(f), S^B(f))$ is called the composition of the family if and only if for $s \in \prod_{x\in B(f)} E_{pr_2 x}^{pr_1 x}$

$$s \in S^B(f) \iff \bigwedge_{x\in B(f)} pr_{pr_1 x}\, s \in S^{B_{pr_1 x}} .$$

Definition 5.3. An information base (B, S^B) is called functional if and only if there is a mapping $F: pr_I S^B \to pr_O S^B$ so that

$$s \in S^B \iff (pr_x s)_{x\in O} = F((pr_x s)_{x\in I}) .$$

The main problem we consider in this section is the following.

Given a family of functional information bases $(B_i, S^{B_i})_{i\in[1,\ell]}$ and a scheme of composition f compatible with the family. Construct the space of available states $S^B(f)$ for the composition $(B(f), S^B(f))$.

5.2 Variable Schemes and Functional Information Bases. The System of Functional Equations Associated with a Variable Scheme.

We associate a variable scheme and the corresponding system of functional equations with a given family of functional information bases and a scheme of composition.

Let $x = \{x_1, x_2, \ldots, x_n\}$, $y = \{y_1, y_2, \ldots, y_p\}$ be n+p distinct signs and $X = \{X_1, \ldots, X_m\}$ be m different sets. Let $\chi\colon x \cup y \to [1,m]$ be a surjection, that is $\chi(x \cup y) = [1,m]$. By $(x \cup y)^+$ we denote the set of all words in the alphabet $x \cup y$ with different signs. Let $u_j \in (x \cup y)^+$, for $j = 1,2,\ldots,p$ and such that y_j does not enter in u_j. Let, now, $F = \{F_1, F_2, \ldots, F_p\}$ be p mappings

$$F_i\colon X_{\chi \circ pr_1(u_i)} \times X_{\chi \circ pr_2(u_i)} \times \ldots \times X_{\chi \circ pr_{\eta_i}(u_i)} \to X_{\chi(y_i)}$$

where $\eta_i = |u_i|$ is the number of signs in the word u_i. The structural set $V = (x,y,X;\ \chi, u_1, \ldots, u_p, F_1, \ldots, F_p)$ $= (x,y,X;\ \chi, u, F)$ is called a *variable scheme*.

Example 5.2. For $x = \{x_1, x_2, x_3\}$; $y = \{y_1, y_2, y_3, y_4\}$, $X = \{X_1, X_2, X_3, X_4\}$;

$t \in x \cup y$	x_1	x_2	x_3	y_1	y_2	y_3	y_4
$\chi(t)$	1	1	2	1	1	3	4

;

$$u_1 = x_1 x_3 y_2\ ;\quad u_2 = y_1 y_3\ ;\quad u_3 = x_2 x_1 y_1 y_2\ ;\quad u_4 = y_1 y_2 y_3\ ;$$

$$F_1\colon X_1 \times X_2 \times X_1 \to X_1\ ;\quad F_2\colon X_1 \times X_3 \to X_1\ ;$$

$$F_3\colon X_1 \times X_1 \times X_1 \times X_1 \to X_3\ ;\quad F_4\colon X_1 \times X_1 \times X_3 \to X_4$$

the structure $(x,y,X;\ \chi,u,F)$ is a variable scheme.

Definition 5.4. Let $V = (x,y,X;\ \chi,u,F)$ be a variable scheme. The relation

$$\text{for every } i,\ 1 \le i \le n, \text{ we have } x_i \in X_{\chi(x_i)}, \text{ and}$$

$$\text{for every } j,\ 1 \le j \le p, \text{ we have } y_j \in X_{\chi(y_j)}, \text{ and}$$

$$y_j = F_j(pr_1(u_j), pr_2(u_j), \ldots, pr_{\eta_j}(u_j)) \equiv F_j(u_j)$$

is called a system of functional equations for dependent variables $y = \{y_1,\ldots,y_p\}$, independent variables $x = \{x_1,x_2,x_n\}$ associated with the variable scheme V.
A family $f = \{f_1,f_2,\ldots,f_p\}$ of mappings

$$f_i\colon X_{\chi\circ pr_1(v_i)} \times \cdots \times X_{\chi\circ pr_{\eta_i}(v_i)} \to X_{\chi(y_i)} ,$$

where $v_i \in x^+$, $1 \le i \le p$, is called a solution for the above system of functional equations if and only if for every j, $1 \le j \le p$ and for all variables from y entering into u_j, that is, for $y_{i_1} = pr_{i_1}(u_j)$, $y_{i_2} = pr_{i_2}(u_j), \ldots, y_{i_k} = pr_{i_k}(u_j)$ we have

$$f_j(v_j) = F_j(pr_1(u_j), \ldots, f_{i_1}(v_{i_1}), \ldots, f_{i_2}(v_{i_2}),$$
$$\ldots, f_{i_k}(v_{i_k}), \ldots, pr_{\eta_j}(u_j))$$

(here, if $i_1 = 1$, then $pr_1(u_j)$ does not enter, and if $i_k = \eta_j$, then $pr_{\eta_j}(u_j)$ does not enter).

Example 5.3. For the variable scheme from example 5.2, we have the following system of equations

$$y_1 = F_1(x_1,x_3,y_2)\ ,\quad y_2 = F_2(y_1,y_3)\ ,$$

$$y_3 = F_3(x_2,x_1,y_1,y_2)\ ,\quad y_4 = F_4(y_1,y_2,y_3) \quad .$$

Now, we assign a variable scheme to a given family $(B_i, S^{B_i})_{i\in[1,\ell]}$ of functional information bases. We will use the following notations.

Let I_i and O_i be the sets of inputs and outputs of i information base. Let

$$I_i = \{I_1^i, I_2^i, \ldots, I_{n_i}^i\}, \quad O_i = \{O_1^i, O_2^i, \ldots, O_{m_i}^i\} .$$

Let F^i

$$F^i\colon E_{I_1^i} \times E_{I_2^i} \times \ldots \times E_{I_{n_i}^i} \to E_{O_1^i} \times E_{O_2^i} \times \ldots \times E_{O_{m_i}^i}$$

be a mapping realizing the transformation from input into output of the base. We denote also

$$F_j^i\colon E_{I_1^i} \times \ldots \times E_{I_{n_i}^i} \to E_{O_j^i} \quad \text{for} \quad 1 \leq j \leq m_i .$$

Now let $X = \{X_1, X_2, \ldots, X_m\}$ be the set of all different sets among all sets $E_{I_j^i}$ and $E_{O_t^i}$ for all $i \in [1,\ell]$ and all $1 \leq j \leq n_i$, $1 \leq t \leq m_i$. Therefore, we have the following functions

$$\alpha^i\colon [1,n_i] \to [1,m] , \quad \beta^i\colon [1,m_i] \to [1,m]$$

for $i \in [1,\ell]$ so that

$$F^i\colon X_{\alpha^i(1)} \times \ldots \times X_{\alpha^i(n_i)} \to X_{\beta^i(1)} \times \ldots \times X_{\beta^i(m_i)}$$

or

$$F_j^i\colon X_{\alpha^i(1)} \times \ldots \times X_{\alpha^i(n_i)} \to X_{\beta^i(j)} .$$

Let

$$\overline{O} = \coprod_{i=1}^{\ell} O_i \ , \quad \overline{I} = \coprod_{i=1}^{\ell} I_i \ , \quad C_2 \subset \overline{O} \ , \quad f\colon C_2 \to \overline{I}$$

be a composition schems. $I = \overline{I} - f(c_2)$. We assume that n_O and n_I are enumerations on $\overline{O}$ and I, that is, we have bejections

$$n_O\colon [1,p] \to \coprod_{i=1}^{\ell} O_i \ , \quad n_O(j) = (i, O_k^i) \quad ,$$

where i and j are functions of i,

$$n_I\colon [1,n] \to \coprod_{i=1}^{\ell} I_i - f(c_2) \ , \quad n_I(j) = (i, I_k^i) \quad ,$$

where i and k are functions of j.

Consider two sets $x = \{x_1, x_2, \dots, x_n\}$, $y = \{y_1, y_2, \dots, y_p\}$ of distinct signs. Now, we can construct the function

$$\chi\colon x \cup y \to [1,m]$$

so that $\chi(x_i) = \alpha_j(k)$, where $n_I(i) = (j, I_k^j)$, and $\chi(y_i) = \beta_j(k)$, where $n_O(i) = (j, O_k^j)$. We construct, now, words $u_j^i \in (x \cup y)^+$ for $i = 1,2,\dots,\ell$, $j = 1,2,\dots,m_i$ so that the number of words $\sum_{i=1}^{\ell} m_i = p$.

We have

$$u^i = u_1^i = u_2^i = \dots = u_{m_i}^i \quad .$$

Let now for $i = 1,2,\dots,\ell$

$$\overline{I}^i = \{(i, I_1^i), (i, I_2^i), \dots, (i, I_{n_i}^i)\} \ , \quad \overline{I} = \bigcup_{i \in [1,\ell]} \overline{I}^i$$

and

$$f(c_2) \cap \overline{I}^i = \{(i, I_{k_1,i}^i), (i, I_{k_2,i}^i), \dots, (i, I_{k_{p_i},i}^i)\} \quad .$$

Then

$$u^i = x_{n_I^{-1}(i,I_1^i)} \cdots x_{n_I^{-1}(i,I^i_{k_1-1,i})} y_{n_O^{-1}(i,I^i_{k_1,i})} x_{n_I^{-1}(i,I^i_{k_1+1,i})}$$

$$\cdots x_{n_I^{-1}(i,I^i_{k_{p_i}-1,i})} y_{n_O^{-1}(i,I^i_{k_{p_i},i})} x_{n_I^{-1}(i,I^i_{k_{p_i}+1,i})}$$

$$\cdots x_{n_I^{-1}(i,I^i_{n_i})} .$$

Therefore, we have p words $u_j \in (x \cup y)^+$, $j = 1,2,\ldots,p$, and p functions $F_1,F_2,\ldots,F_p$,

$$u_j = u^i_k , \quad F_j = F^i_k ,$$

where $n_O(j) = (i,O^i_k)$. Therefore for a given family $(B_i,S^{B_i})_{i\in[1,\ell]}$ of functional information bases we associate the variable scheme $V = (x,y,X; \chi,u,F)$. We call this variable scheme *a canonical variable scheme associated with a family of functional information bases and a scheme of composition* (or, simpler, a *canonical variable scheme*).

We consider now primitive functional information bases. An element base $B = (A,C,I,O; \partial_O,\partial_I)$ is called a *primitive element base* if the set A contains only one element, say a, that is, $A = \{a\}$, and the set C is the empty set, that is, $C = \emptyset$. An information base $IB = (B,S^B)$ is called a *primitive information base* if its element base B is a primitive element base, and the space of states of its unique element contains only one element, say ℓa; that is, $E_a = \{\ell a\}$. The space of available states S^B of a primitive information base is determined by spaces of states of its inputs and outputs. Therefore, we will identify the space of available states S^B of a primitive information base (B, S^B) with its projection on inputs and outputs, that is, with the set $pr_{I \cup O}\, S^B$. By abusing language, we will call this projection as a space of available states. Therefore,

$$S^B \subset \prod_{x \in I \cup O} E_x$$

where E_x is the space of states of the element x.

Theorem 5.1. *For the composition* $(B(f), S^B(f))$ *of a family* $(B_i, S^{B_i})_{i \in [1,\ell]}$ *of primitive functional information bases and a scheme of composition* f *compatible with the family the space* $S^B(f)$ *is a set of solutions of the system of equations determined by the canonical variable scheme.*

The analogous theorem may be proved for a general functional information base. We are not going to formulate this theorem here.

5.3 Scheme-Vector Functions

In this point we develop a technique which will permit us to construct solutions for systems of functional equations associated with variable schemes.

Example 5.4. We will use functions of the following type, for example,

$$F\colon X_2 \times X_1 \times X_1 \times X_2 \to Y_1 \times Y_1 \times Y_2\colon \quad (x_1, x_2, x_3, x_4)$$

$$\to (y_1, y_2, y_3) = (f_1(x_3, x_1, x_2), f_2(x_1, x_3), f_3(x_4, x_3)).$$

and will compose such functions. The defined abouve function F may be described by the graph

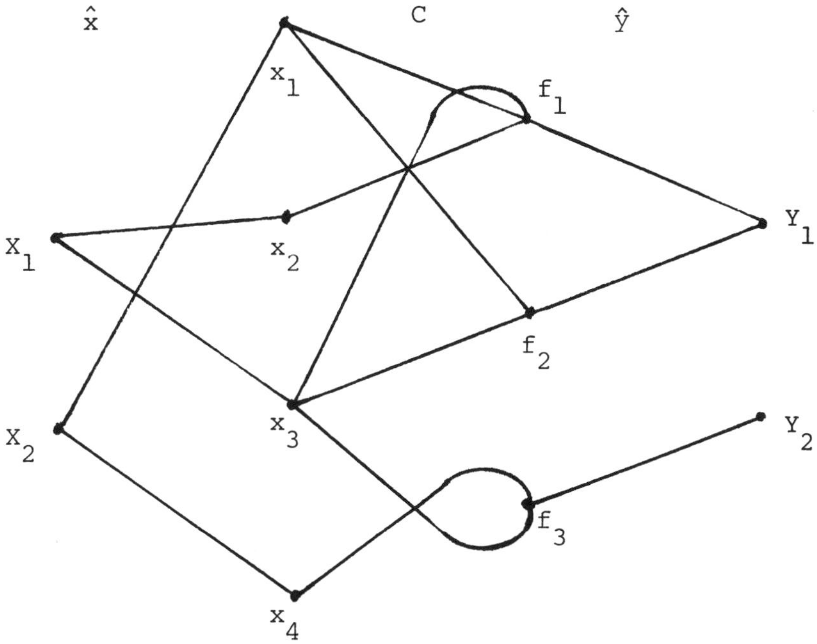

This function is determined by sets from $X = \{X_1, X_2\}$ and from $Y = \{Y_1, Y_2\}$; by functions from $f = \{f_1, f_2, f_3\}$, where $f_1 \colon X_1 \times X_2 \times X_3 \to Y_1$, $f_2 \colon X_2 \times X_1 \to Y_1$, $f_3 \colon X_2 \times X_1 \to Y_2$; by two mappings $\hat{x} \colon \{1,2,3,4\} \to \{1,2\}$, where $\hat{x}(1) = \hat{x}(4) = 2$, $\hat{x}(2) = \hat{x}(3) = 1$, and $\hat{y} \colon \{1,2,3\} \to \{1,2\}$, where $\hat{y}(1) = \hat{y}(2) = 1$, $\hat{y}(3) = 2$; and by the graph C between two layers $x = \{x_1, x_2, x_3, x_4\}$ and $f = \{f_1, f_2, f_3\}$. It should be noted that arcs entering the vertices f_1, or f_2, or f_3 are totally ordered.

In a general situation, we have ℓ different sets from $X = \{X_1, X_2, \ldots, X_\ell\}$ and p different sets from $Y = \{Y_1, Y_2, \ldots, Y_p\}$; some of X_i and Y_j may be the same. We have also two surjections $\hat{x} \colon [1,n] \to [1,\ell]$ and $\hat{y} \colon [1,m] \to [1,p]$, and m words from $u = \{u_1, u_2, \ldots, u_m\}$ in the alphabet $[1,n]$ with different signs, that is, $u_i \in [1,n]^+$ for $i = 1,2,\ldots,m$.

Let $f = \{f_1,f_2,\ldots,f_m\}$ be m mappings: for $i = 1,2,\ldots,m$

$$f_i\colon X_{\hat{x}(pr_1(u_i))} \times X_{\hat{x}(pr_2(u_i))} \times \ldots \times U_{\hat{x}(pr_{\eta_i}(u_i))} \to Y_{\hat{y}(i)} \;.$$

Let $C(n,m,u)$ be a graph between two layers: the first layer is the set $[1,n]$, the second layer is the set $[1,m]$; and for every $i \in [1,n]$ and $j \in [1,m]$ there is an arc from i into j if and only if i enters into the word u_j. Every word u_j being an ordered set of symbols from $[1,n]$ induces naturally the totoal order on the set of arcs entering the vertex j. This graph will be called a *scheme-graph*. Therefore, we have a structure $(X,Y;\ \hat{x},\hat{y},f,C(n,m,u))$ which will be called a *function scheme*.

Definition 5.5. Given a function-scheme $(X,Y;\ \hat{x},\hat{y},f,$ $C(n,m,u))$. The function

$$F\colon X_{\hat{x}(1)} \times X_{\hat{x}(2)} \times \ldots \times X_{\hat{x}(n)} \to Y_{\hat{y}(1)} \times Y_{\hat{y}(2)} \times \ldots \times Y_{\hat{y}(m)}\colon$$

$$(x_1,x_2,\ldots,x_n) \to (y_1,y_2,\ldots,y_m)$$

$$= (f_1(\overline{x}(u_1)),f_2(\overline{x}(\overline{u}_2)),\ldots,f_m(\overline{x}(u_m)) \quad ,$$

where $\overline{x}(u_i) = (x_{pr_1(u_i)},x_{pr_2(u_i)},\ldots,pr_{\eta_i}(u_i))$, is called a primitive scheme-vector function associated with the function-scheme $(X,Y\colon \hat{x},\hat{y},f,c(n,m,u))$.

If we have another function-scheme $(Y,Z;\ \hat{y},\hat{z},\phi,D(m,p,v))$ and therefore, the corresponding primitive scheme-vector function

$$\phi\colon Y_{\hat{y}(1)} \times Y_{\hat{y}(2)} \times \ldots \times Y_{\hat{y}(m)} \to Z_{\hat{z}(1)} \times Z_{\hat{z}(2)} \times \ldots \times Z_{\hat{z}(p)}:$$

$$(y_1, y_2, \ldots, y_m) \to (z_1, z_2, \ldots, z_p)$$

$$= (\phi_1(\bar{y}(v_1)), \phi_2(\bar{y}(v_2)), \ldots, \phi_p(\bar{y}(v_p))$$

then we define the composition of those two scheme-vector functions as follows:

$$(\phi \circ F)_{D\circ C}\colon X_{\hat{x}(1)} \times X_{\hat{x}(2)} \times \ldots \times X_{\hat{x}(n)} \to Z_{\hat{z}(1)} \times Z_{\hat{z}(2)} \times \ldots \times Z_{\hat{z}(p)}:$$

$$(x_1, x_2, \ldots, x_n) \to (z_1, z_2, \ldots, z_p)$$

$$= \Big(\phi_1\Big(f_{pr_1(v_1)}(\bar{x}(u_{pr_1(v_1)})),\ f_{pr_2(v_1)}(\bar{x}(u_{pr_2(v_1)})),$$

$$\ldots,\ f_{pr_{|v_1|}(v_1)}(\bar{x}(u_{pr_{|v_1|}(v_1)}))\Big),$$

$$\phi_2\Big(f_{pr_1(v_2)}(\bar{x}(u_{pr_1(v_2)})),\ f_{pr_2(v_2)}(\bar{x}(u_{pr_2(v_2)})),$$

$$\ldots,\ f_{pr_{|v_2|}(v_2)}(\bar{x}(u_{pr_{|v_2|}(v_2)}))\Big),$$

$$\ldots \quad \ldots \quad \ldots \quad \ldots$$

$$\phi_p\Big(f_{pr_1(v_p)}(\bar{x}(u_{pr_1(v_p)})),\ f_{pr_2(v_p)}(\bar{x}(u_{pr_2(v_p)})),$$

$$\ldots,\ f_{pr_{|v_p|}(v_p)}(\bar{x}(u_{pr_{|v_p|}(v_p)}))\Big)\Big) \quad .$$

Here we used the notation: $|v_i|$ is the number of signs in the word v_i. We denote the scheme of this function as $(X,Z;\ \hat{x},\hat{z},(f,\phi),D\circ C)$. The graph $D\circ C$ is called a *composition graph*.

Example 5.5. Let ϕ be a primitive scheme-vector function

$$\phi\colon Y_1 \times Y_1 \times Y_2 \to Z_2 \times Z_1\colon (y_1,y_2,y_3) \to (z_1,z_2)$$

$$= (\phi_1(y_1,y_3),\phi_2(y_2,y_1))$$

with the scheme

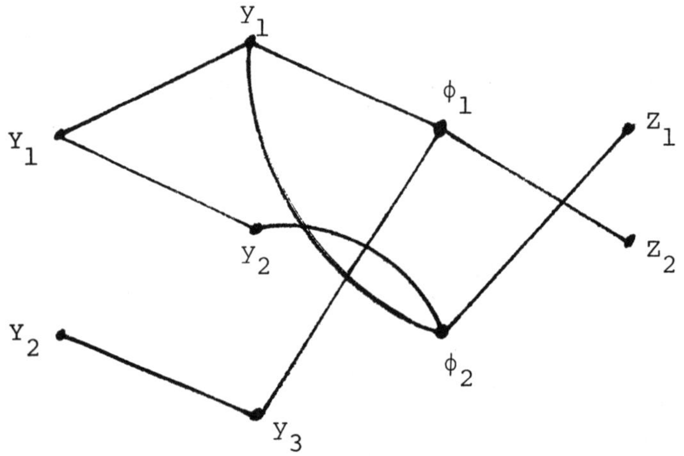

Then the composition of the function F from the example 5.4 and the function ϕ have the following scheme-graph

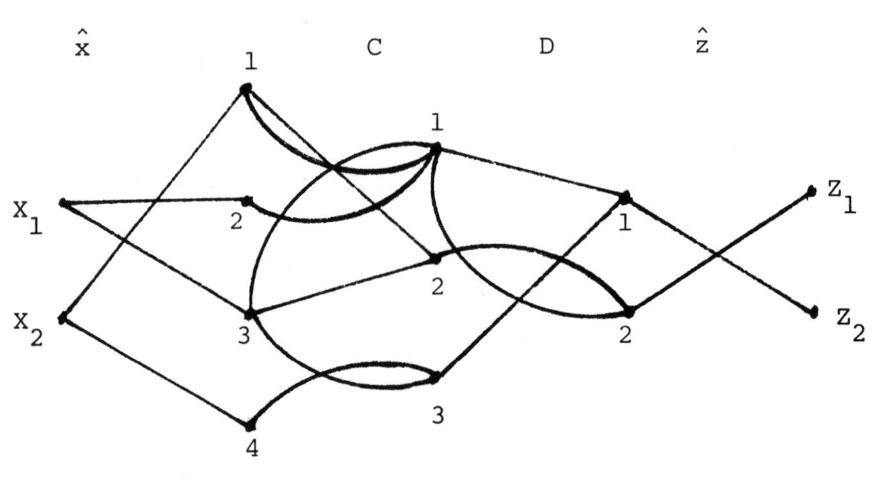

The composition

$$(\phi \circ F)_{D\circ C}: X_2 \times X_1 \times X_1 \times X_2 \to Z_2 \times Z_1: (x_1,x_2,x_3,x_4) \to (z_1,z_2)$$

$$= \big(\phi_1(f_1(x_3,x_1,x_2),f_3(x_4,x_3)),\phi_2(f_2(x_1,x_3),f_1(x_3,x_1,x_2))\big).$$

Let us consider the class of all primitive scheme-vector functions and all possible compositions of these primitive scheme-vector functions. Every element of this class is associated with the scheme $(X,Y; \hat{x},\hat{y},(f^1,f^2,\ldots,f^p),\Gamma)$, where $f^1,f^2,\ldots,f^p$ are vector-functions of the second, third,...,$p+1^{th}$ layer of the composition graph Γ. Composition graph Γ is a parallel-layer graph, that is, the set A of vertices of the graph comprises layers $A_1,A_2,\ldots,A_{p+1}$ such that $A_i \cap A_j = \emptyset$, and $A = \cup_{i=1}^{p+1} A_i$, and all arcs are from the previous layer to the next one; that is, if c is an arc and $pr_1(c)$ and $pr_2(c)$ are the beginning and the end of the arc, $pr_1(c) \in A_i$, $pr_2(c) \in A_j$, then $j = i+1$. But we can extend now the set of composition graphs to layer graphs, that is, to graphs $\Gamma = (A,C)$, $C \subset A \times A$, $A = \cup_{i=1}^{p+1} A_i$, $A_i \cap A_j = \emptyset$, and if $c \in C$, $pr_1(c) \in A_i$, $pr_2(c) \in A_j$, then $j > i$. Really, for every arc from layer i to layer j $(j > i)$, we can introduce fictitious vertices on layers $i+1,i+2,\ldots,i+k = j-1$ with one input and one output and identity function. The example below clarifies this construction.

Example 5.6.

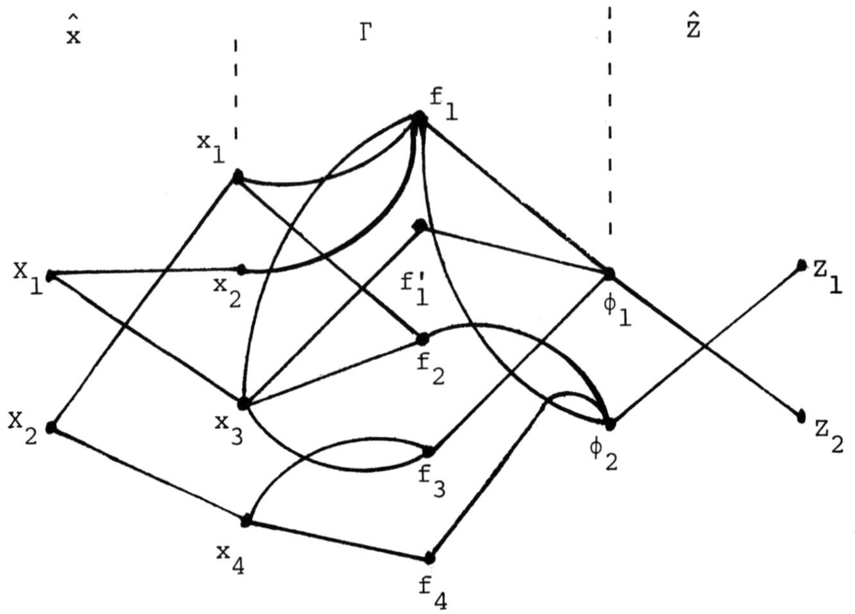

We have the following function scheme $(X,Z;\ \hat{x},\hat{z},(f,\phi),\Gamma)$, $X = \{X_1,X_2\}$, $Z = \{Z_1,Z_2\}$, $f = \{f_1,f_2,f_3\}$, $\phi = \{\phi_1,\phi_2\}$, $\hat{x}$, $\hat{z}$, and Γ are depicted on the figure above. Γ is a layer graph. We have

$$f_1\colon X_1 \times X_2 \times X_1 \to Y_1\ , \quad f_2\colon X_2 \times X_1 \to Y_2,$$

$$f_3\colon X_2 \times X_1 \to Y_3 \quad ,$$

$$\phi_1\colon Y_1 \times X_1 \times Y_3 \to Z_2\ , \quad \phi_2\colon Y_2 \times X_2 \times X_1 \to Z_1$$

The corresponding vector-function

$$T\colon X_2 \times X_1 \times X_1 \times X_2 \to Z_2 \times Z_1\colon (x_1,x_2,x_3,x_4) \to (z_1,z_2)$$

$$= \big(\phi_1(f_1(x_3,x_1,x_2),x_3,f_3(x_4,x_3)),$$

$$\phi_2(f_2(x_1,x_3),x_4,f_1(x_3,x_1,x_2))\big) \quad .$$

If we introduce two new functions

$$f_1: X_1 \to X_1 , \quad f_1 = 1_{X_1} ,$$

$$f_4: X_2 \to X_2 , \quad f_4 = 1_{X_2} ,$$

then we receive two primitive vector-scheme functions F and ϕ associated correspondingly with the functional scheme $(X,Y;\ \hat{x},\hat{y},f',\Gamma_1)$, where $Y = \{Y_1,X_1,Y_2,Y_3,X_2\}$, $f' = \{f_1,f_1',f_2,f_3,f_4\}$, and $\hat{x}$, $\hat{y}$, Γ_1 are depicted on the figure; and the functional scheme $(Y,Z;\ \hat{y},\hat{z},\phi,\Gamma_2)$, where Γ_2 is depicted on the figure. Now, the function T is a composition of function F and ϕ, that is, $T = (\phi \circ F)_{\Gamma_2 \circ \Gamma_1}$.

Definition 5.6. A scheme-vector function is a vector-function associated with a functional vector scheme of the type $(X,Y;\ \hat{x},\hat{y},(f_1,f_2,\ldots f_p),\Gamma)$, where: Γ is a layer graph with p+1 layers; $f_1,f_2,\ldots,f_p$ are vector-functions comprising the second, the third,..., and the p+1th layer of the graph Γ; for every vertex of the layers $f_1,\ldots,f_p$, the set of input arcs is totally ordered set. A scheme-vector function associated with a functional vector scheme will be denoted often by this scheme.

5.4 Solution of a System of Functional Equations Associated with a Variable Scheme

Definition 5.7. Given a variable scheme $V = (x,y,X;\ \chi,u,F)$. The graph Γ_V with vertices $x \cup y$ and arcs $(a_i,y_j) \in \Gamma_V$ if and only if a_i is a sign of the word u_j is called *associated with the variable scheme* V.

We consider first the case where the graph Γ_V does not contain cycles. We represent such a graph in canonical form.

Namely, we use

Lemma 5.1. *If* $\Gamma = (A,C)$, *where* $C \subset A \times A$, *is a graph without cycles, then there exists a partition* $(A_i)_{i \in [0,\ell]}$ *of the set of vertices* A, *that is,* $A = \cup_{i \in [0,\ell]} A_i$, $A_i \cap A_j = \emptyset$ *if* $i \neq j$, *such that for every* $c \in C$, *if* $pr_1(c) \in A_i$ and $pr_2(c) \in A_j$, then $i < j$.

Sets A_i for $i = 0,1,\ldots,\ell$ are called layers of the graph. The partition $(A_i)_{i \in [0,\ell]}$ is called a *canonical form of the graph* $\Gamma = (A,C)$. If Γ_V is the graph associated with a variable scheme $V(x,y,X;\ \chi,u,F)$ and $\{v_0,v_1,\ldots,v_r\}$ is a canonical form of Γ, then it is obvious that $v_0 = x$ and for every $x_i \in x$ we have $x_i \notin v_j$ for $j = 1,2,\ldots,r$, that is, $v_1 \cup v_2 \cup \ldots \cup v_r = y$. We have, therefore, a partition $(v_i)_{i \in [1,r]}$ of the set y.

We will use the following notations. If a set $X = \{x_1,x_2,\ldots,x_k\}$ is totally ordered, for example, $x_1 < x_2 < \ldots < x_k$, then we denote this set as $<x_1,x_2,\ldots,x_k>$. We will use the function $p_X\colon X \to [1,k]$, where $p_X(x_i) = i$, that is, $p_X(x_i)$ is the position of x_i in the ordered set x.

Definition 5.8. Given a variable scheme $V = (x,y,X;\ \chi,u,F)$, which associated graph Γ_V does not contain cycles. We assume $x = \{x_1,x_2,\ldots,x_n\}$; $y = \{y_1,y_2,\ldots,y_p\}$; $X = \{X_1,X_2,\ldots,X_\ell\}$; $\chi\colon x \cup y \to [1,m]$ and χ is a surjection; $u = \{u_1,u_2,\ldots,u_p\}$ and $u_i \in (x \cup y)^+$ for $i = 1,\ldots,p$ are words in alphabet $x \cup y$ with different signs, y_i does not enter u_i; $F = \{F_1,F_2,\ldots,F_p\}$. The family of scheme-vector functions $(\bar{X},Y_i;\ \hat{x},\hat{y}_i,(f^1,f^2,\ldots,f^i),\Gamma_i)_{i \in [1,r]}$ is called associated with the variable scheme if

i) $\overline{X} = \cup_{i\in[1,m]} \{X_{\chi(x_i)}\}$; we assume that $\overline{X}$ is totally ordered $\overline{X} = \langle X_{i_1}, X_{i_2}, \ldots, X_{i_m}\rangle$;

ii) if $\langle v_0, v_1, \ldots, v_r\rangle$ is a canonical form of Γ_V, then $Y_i = \cup_{y\in V_i} \{X_{\chi(y)}\}$; we assume that V_i and Y_i are totally ordered, $V_i = \langle y_{k_{i1}}, y_{k_{i2}}, \ldots, y_{k_{im}}\rangle$, $Y_i = \langle X_{i1}, X_{i2}, \ldots, X_{i\ell_i}\rangle$;

iii) $\hat{x}$: $[1,n] \to [1,m]$: $t \to \hat{x}(t) = p_{\overline{X}}(X_{\chi(x_t)})$, $\hat{x}$ is a surjection;

iv) $\hat{y}_i$: $[1,m_i] \to [1,\ell_i]$: $t \to \hat{y}_i(t) = p_{Y_i}(X_{\chi(y_{k_{it}})})$;

v) $f^1 = \langle F_{k_{11}}, F_{k_{12}}, \ldots, F_{k_{1m_1}}\rangle$,

$f^2 = \langle F_{k_{21}}, F_{k_{22}}, \ldots, F_{k_{2m_2}}\rangle, \ldots,$

$f^i = \langle F_{k_{i1}}, F_{k_{i2}}, \ldots, F_{k_{im_i}}\rangle$;

vi) Γ_i is the graph with the layers $\langle v_0, f^1, f^2, \ldots, f^i\rangle$; an arc $(a_j, b_t) \in \Gamma_i$ if and only if for $b_t = F_{k_{p_t q_t}}$ and $a_j = F_{k_{p_j q_j}}$ or $a_j = x_{p_j}$, the variable $y_{k_{p_j q_j}}$ or x_{p_j} enters into the word $u_{k_{p_t q_t}}$. The order on the set of arcs entering a vertex $F_{k_{pq}}$ is determined by the word $u_{k_{pq}}$.

Example 5.7. Let for variable scheme $V = (x, y, X; \chi, u, F)$, $x = \{x_1, x_2, x_3, x_4\}$; $y = \{y_1, y_2, y_3, y_4, y_5\}$; and $X = \{X_1, X_2, X_3, X_4, X_5, X_6, X_7\}$;

$x \cup y$	x_1	x_2	x_3	x_4	y_1	y_2	y_3	y_4	y_5
$\chi(t)$	2	1	1	2	3	4	5	7	6

$$u = \{u_1, u_2, u_3, u_4, u_5),$$

where

$$u_1 = x_3x_1x_2 , \quad u_2 = x_1x_3 , \quad u_3 = x_4x_3 , \quad u_4 = y_1x_3y_3 ,$$
$$u_5 = y_2x_4y_1$$

$$F_1: X_1 \times X_2 \times X_1 \to X_3 , \quad F_2: X_2 \times X_1 \to X_4 , \quad F_3: X_2 \times X_1 \to X_5 ,$$
$$F_4: X_3 \times X_1 \times X_5 \to X_7 , \quad F_5: X_4 \times X_2 \times X_3 \to X_6$$

For the family of scheme-vector functions $\{(\bar{X}, Y_1; \hat{x}, \hat{y}_1, f^1, \Gamma_1), (\bar{X}, Y_2; \hat{x}, \hat{y}_2, (f^1, f^2), \Gamma_2)\}$ associated with the variable scheme we have

$$\bar{X} = \langle X_1, X_2\rangle , \quad V_1 = \langle y_1, y_2, y_3\rangle , \quad V_2 = \langle y_4, y_5\rangle$$
$$Y_1 = \langle X_3, X_4, X_5\rangle , \quad Y_2 = \langle X_6, X_7\rangle$$
$$\hat{x}: [1,4] \to [1,2] , \quad \hat{x}(1) = \hat{x}(4) = 2 , \quad \hat{x}(2) = \hat{x}(3) = 1$$
$$\hat{y}_1: [1,3] \to [1,3] , \quad \hat{y}_1(1) = 3 , \quad \hat{y}_1(2) = 2 , \quad \hat{y}_1(3) = 3,$$
$$\hat{y}_2: [1,2] \to [1,2] , \quad \hat{y}_2(1) = 2 , \quad \hat{y}_2(2) = 1$$
$$f^1 = \langle F_1, F_2, F_3\rangle , \quad f^2 = \langle F_6, F_7\rangle .$$

For the scheme-vector function $(\bar{X}, Y_1; \hat{x}, \hat{y}_1, f^1, \Gamma_1)$ we have the scheme

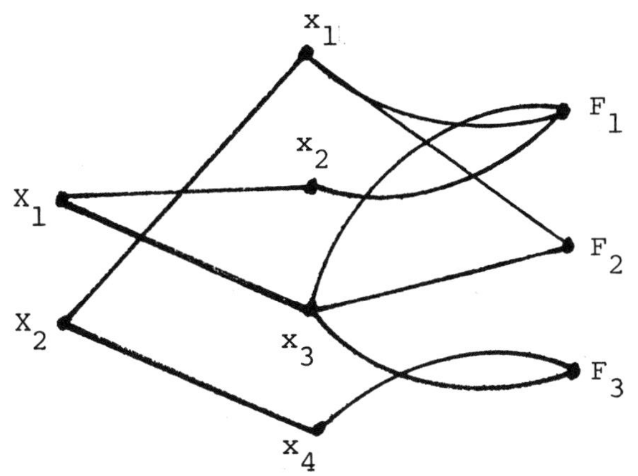

The scheme for the scheme-vector function $(\overline{X}, Y_2;\ \hat{x}, \hat{y}_2, (f^1, f^2), \Gamma_2)$ is the same as in example 5.6, where $f_1 = F_1$, $f_2 = F_2$, $f_3 = F_3$, $\phi_1 = F_4$, $\phi_2 = F_5$, $\hat{y}_2 = \hat{z}$.

Theorem 5.2. *For the system of functional equations associated with a variable scheme* $V = (x, y, X;\ \chi, u, F)$ *with associated graph* Γ_V *without cycles and with the canonical form* $\langle x, v_1, v_2, \ldots, v_r \rangle$, *the family of functions* $(y_i)_{i \in [1,p]}$, *for which* $y_i \in v_q$ *and*

$$y_i = pr_{p_{v_q}(y_i)} \phi_q \quad ,$$

where ϕ_q *is the scheme-vector function from the family* $(\phi_q)_{q \in [1,r]}$ *of scheme-vector functions associated with the variable scheme* V, *is the solution of this system of equations. This solution is unique.*

We now consider the case when the graph Γ_V associated with a variable scheme V has cycles. We will use the notion of bicompoent of a graph. A maximal subgraph of a graph with the following property:

for any two vertices of this subgraph there is a walk between them

is called a *bicomponent*.

A graph with cycles has bicomponents. To a given graph Γ with bicomponents we correspond the graph Γ^* as follows. The vertices of Γ^* correspond to bicomponents of Γ. If x_1 and x_2 are different vertices of Γ^*corresponding to the bicomponents Γ_1 and Γ_2, then the arc $(x_1,x_2) \in \Gamma^*$ if and only if there is an arc from a vertex from Γ_1 to a vertex from Γ_2. It is obvious, that a graph Γ^* is a graph without cycles. The canonical form of the graph Γ^* will be called a canonical form for the bicomponents of Γ.

If the graph Γ_V associated with a variable scheme V has cycles, then we consider the graph Γ_V^* of bicomponents of Γ_V. Subsystems of functional equations correspond to bicomponents. If we resolve these subsystems and substitute solutions into the graph Γ_V^*, then the new corresponding system is associated with the graph without cycles. Therfore, the solution of this new system and, hence, the original system may be obtained by the previous method. We demonstrate this technique in the following example.

Example 5.8. Given the variable scheme $V = (\{x\},y_1\{X\};\chi,u,F)$, where $y = \{y_1,y_2,y_3,y_4,y_5,y_6,y_7\}$; $\chi(x) = \chi(y_i) = X$ for $i=1,2,\ldots,7$; $u = \{u_1,u_2,u_3,u_4,u_5,u_6,u_7\}$; $u_1 = y_2y_3$, $u_2 = xy_1$, $u_3 = xy_1$, $u_4 = y_1$, $u_5 = y_1$, $u_6 = y_3$, $u_7 = y_4$; $F = \{F_1,F_2,F_3,F_4,F_5,F_6,F_7\}$, $F_1\colon X^2 \to X$, $F_2\colon X^2 \to X$, $F_3\colon X^2 \to X$, $F_4\colon X \to X$, $F_5\colon X \to X$, $F_6\colon X \to X$, $F_7\colon X \to X$. The corresponding system of functional equations is $y_1 = F_1(y_2,y_3)$, $y_2 = F_2(x_1,y_1)$, $y_3 = F_3(x_1,y_1)$, $y_4 = F_4(y_1)$, $y_5 = F_5(y_1)$, $y_6 = F_6(y_3)$, $y_7 = F_7(y_4)$.

The corresponding graph in canonical form is

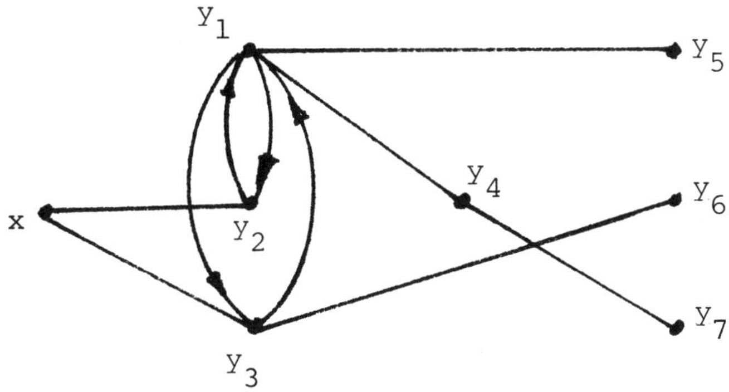

A bicomponent corresponds to the following subsystem

$$y_1 = F_1(y_2, y_3)$$
$$y_2 = F_2(x, y_1)$$
$$y_3 = F_3(x, y_1) \quad .$$

If $y_1 = f_1(x)$, $y_2 = f_2(x)$, and $y_3 = f_3(x)$ is a solution of this subsystem, then the variable scheme for the transformed system of equations may be depicted as

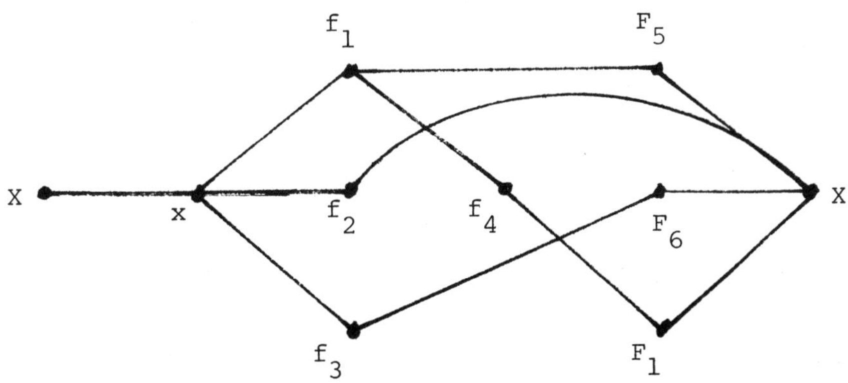

Therefore, the solution of the original system is $y_4 = F_4(f_1(x))$, $y_5 = F_5(f_1(x))$, $y_6 = F_6(f_3(x))$, $y_7 = F_7(F_4(f_1(x)))$. Let, for example, $X = C_0^\infty([0,\infty))$, that is, the space of infinity differential functions on $[0,\infty)$ such that $y \in X$ follows $y(0) = 0$. Let

$$\begin{cases} y_1 = ay_2 + by_3 \\ y_2 = cx + d\dot{y} \\ y_3 = ex + f\dot{y} \end{cases}, \quad \Delta = ad + bf \neq 0$$

where $\dot{y}\colon [0,\infty) \to \mathbf{R}\colon t \to \frac{d}{dt}\, y(t)$. Then

$$y_1\colon [0,\infty) \to \mathbf{R}\colon t \to y_1(t)$$

$$= -(ac+be)/\Delta e^{t/\Delta} \int_0^t e^{-\tau/\Delta}\, x(\tau)\, d\tau ,$$

$$y_2\colon [0,\infty) \to \mathbf{R}\colon t \to y_2(t)$$

$$= \frac{b(cf-de)}{\Delta}\, x(t) - \frac{d(ac+be)}{\Delta^2}\, e^{t/\Delta} \int_0^t e^{-\tau/\Delta}\, x(\tau)\, d\tau ,$$

$$y_3\colon [0,\infty) \to \mathbf{R}\colon t \to y_3(t)$$

$$= \frac{-a(cf-de)}{\Delta}\, x(t) - \frac{f(ac+be)}{\Delta^2}\, e^{t/\Delta} \int_0^t e^{-\tau/\Delta}\, x(\tau)\, d\tau ,$$

$$y_4 = F_4(y_1), \quad y_5 = F_5(y_1), \quad y_6 = F_6(y_3),$$

$$y_7 = F_7(F_4(y_1)) .$$

6. CATEGORY OF SPACES OF AVAILABLE STATES AND ITS APPLICATION TO MODELING COMPUTERS AND THEIR LANGUAGES

In this section, I consider another particular case of information bases called spaces of available states. This type of information base is used for modeling subsystems of a computer and its instructions. Some constructions introduced are interpreted in terms of the theory of categories.

6.1 Category of Spaces of Available States

Definition 6.1. A structure $(X, \&_X; E_X)$, where $E_X : X \to \beta(\&_X)$ is a function from X into $\beta(X)$, such that $\cup_{y \in X} E_X(y) = \&_X$ is called a space of states for elements from X. The set $E^y = E_X(y) \subset \&_X$ for $y \in X$ is called a space of states of the element y. A space of states $(X, \&_X; E_X)$ is a distribution of the set $\&_X$ among elements from X, defined in point 1.1.

From the intuitive point of view, a space of states for elements from X defines "the physical meaning" to the elements from X. The set X "assigns labels" to elements, and the set $E_X(y)$ for a given element y defines the "physical meaning" for this element (its possible states).

By using the technique developed in section 2, we can define canonical morphisms for the structure of spaces of states.

Proposition 6.1. *Given two spaces of states* $\alpha_1 = (X_1, \&_{X_1}; E_{X_1})$ *and* $\alpha_2 = (X_2, \&_{X_2}; E_{X_2})$. *A pair of mappings* $(f_X, f_{\&_X})$

$$f_X : X_1 \to X_2 , \quad f_{\&_X} : \&_{X_1} \to \&_{X_2} ,$$

which hold the condition

$$\beta(f_X \times \beta(f_{\mathcal{E}_X}))(E_{X_1}) \subset E_{X_2}$$

is a canonical morphism from the structure α_1 *into the structure* α_2. *A pair* $(f_X, f_{\mathcal{E}_X})$ *is a canonical morphism if and only if the diagram*

$$\begin{array}{ccc} X_1 & \xrightarrow{\;f_X\;} & X_2 \\ {\scriptstyle E_{X_1}}\downarrow & & \downarrow{\scriptstyle E_{X_2}} \\ \beta(\mathcal{E}_{X_1}) & \xrightarrow{\;\beta(f_{\mathcal{E}_X})\;} & \beta(\mathcal{E}_{X_2}) \end{array}$$

is commutative.

We are going to introduce the coproduct for two given spaces of states. In general, a family of morphisms $(\alpha_i : A \to A_i)_{i \in I}$ from the object A into the objects A_i of a given category J is called a *cone with the initial object* A. A family of morphisms $(\sigma_i : A_i \to A)_{i \in I}$ from the objects A_i into the object A is called a *cocone with the final object* A. Given a family $(A_i)_{i \in I}$ of objects of a category J. An object A of the category J and a cocone $(\sigma_i : A_i \to A)_{i \in I}$ with the final object A are called the *coproduct of objects* A_i, $i \in I$, if for any cocone $(\alpha_i : A_i \to B)$ there exists the unique morphism $\alpha : A \to B$ such that $\alpha_i = \alpha\sigma_i$ for every $i \in I$ (here $\alpha\sigma_i$ is the composition of morphisms α and σ_i in the category J).

Proposition 6.2. *Let* $(\alpha_i)_{i \in I} = (X_i, \mathcal{E}_{X_i}; E_{X_i})_{i \in I}$ *be a family of spaces of states. The space of states*

$$\alpha = \left(\coprod_{i \in I} X_i \,,\; \coprod_{i \in I} \mathcal{E}_{X_i} \,;\; E_{\amalg X_i} \right)$$

where

$$E_{\amalg X_i} : \coprod_{i\in I} X_i \to \beta\left(\coprod_{i\in I} \mathcal{E}_{X_i}\right) : (i,x) \to \{i\} \times E_{X_i}(x)$$

and morphisms

$$\sigma_i : (X_i, \mathcal{E}_{X_i}; E_{X_i}) \to \alpha$$

where $\sigma_i = (\phi_{X_i}, \phi_{\mathcal{E}_{X_i}})$,

$$\phi_{X_i} : X_i \to \coprod_{i\in I} X_i : x \to \phi_{X_i}(x) = (i,x)$$

$$\phi_{\mathcal{E}_{X_i}} : \mathcal{E}_{X_i} \to \coprod_{i\in I} \mathcal{E}_{X_i} : e \to \phi_{\mathcal{E}_{X_i}}(e) = (i,x)$$

are the coproduct of the spaces α_i.

The coproduct of the family $(\alpha_i)_{i\in I}$ is denoted by $\coprod_{i\in I} \alpha_i$. The "space of available states" as structure $S^X = (X, \mathcal{E}_X; E_X, S_X)$, where $(X, \mathcal{E}_X; E_X)$ is the space of states and S_X is a set of available states $S_X \subset \prod_{y\in X} E_X(y)$, was introduced in the point 4.1. Therefore, the structure (X, S^X) is an information base, where the element base X is a set, that is, the sets of connections, inputs, and outputs are empty sets. In the point 4.1, we described morphisms from a space of available states $(X_1, \mathcal{E}_{X_1}; E_{X_1}, S_{X_1})$ into a space of available states $(X_2, \mathcal{E}_{X_2}; E_{X_2}, S_{X_2})$. Here we define coproducts in the category of spaces of available states.

Proposition 6.3. *Let* $(\eta_i)_{i\in I} = (X_i, \mathcal{E}_{X_i}; E_{X_i}, S_{X_i})_{i\in I}$ *be a family of spaces of available states. The space of available states*

$$\eta = \coprod_{i\in I} \eta_i = \left(\coprod_{i\in I} X_i , \coprod_{i\in I} \&_{X_i} ; E_{\amalg X_i} , \square_{i\in I} S_{X_i}\right) ,$$

where

$$s \in \square_{i\in I} S_{X_i} \Longleftrightarrow \left\{ s: \coprod_{i\in I} X_i \to \coprod_{i\in I} \&_{X_i} , \bigwedge_{i\in I} \bigvee_{s_i \in S_i} s(i,x) = (i, s_i(x)) \right\} ,$$

and morphisms $\sigma_i : \eta_i \to \eta$, *such that for every* $i \in I$ σ_i *are morphisms from* $\alpha_i = (X_i, \&_{X_i} ; E_{X_i})$ *into* $\alpha = \coprod_{i\in I} \alpha_i$ *defined in Proposition 6.2, are the coproduct of the family* $(\eta_i)_{i\in I}$ *of spaces of available states.*

If sets of the family $(X_i)_{i\in I}$ are disjoint, that is, $X_i \cap X_j = \emptyset$ for $i \neq j$, then we will use the set $\cup_{i\in I} X_i$ instead of $\coprod_{i\in I} X_i$. In this case, we have a bijection

$$\Delta: \bigcup_{i\in I} X_i \to \coprod_{i\in I} X_i : x \to (i,x) \text{ if } x \in X_i ,$$

and the inverse function is pr_2, that is,

$$pr_2 : \coprod_{i\in I} X_i \to \bigcup_{i\in I} X_i : (i,x) \to pr_2(i,x) = x .$$

We will use the function $E_{\cup X_i}$ instead of $E_{\amalg X_i}$, that is,

$$E_{\cup X_i} = E_{\amalg X_i} \Delta : \bigcup_{i\in I} X_i \to \beta\left(\coprod_{i\in I} \&_{X_i}\right) : x \to \{i\} \times E_{X_i}(x)$$

for $x \in X_i$; and the set $\frown_{i\in I} S_{X_i}$, where

$$s \in \bigcap_{i\in I} S_{x_i} \iff (s: \bigcup_{i\in I} X_i \to \coprod_{i\in I} E_{x_i},\ s = \hat{s}\cdot\Delta,$$

$$\hat{s} \in \square_{i\in I} S_{x_i}) \quad ,$$

or formally,

$$\bigcap_{i\in I} S_{x_i} = \square_{i\in I} S_{x_i} \cdot \Delta \quad .$$

The space of available states $\coprod_{i\in I} \eta_i$ will be denoted in this case by $\bigcup_{i\in I} \eta_i$.

6.2 Finite Spaces of (Available) States

6.2.1. Let $\alpha = (X, \&_x; E_x)$ $(\eta = (X, \&_x; E_x, S_x))$ be a space of (available) states. We call this space by a *finite space of (available) states* if the set X is finite, that is, $X = \{x_1, x_2, \ldots, x_n\}$. We have in this case $E_x(x_1) = E_{x_1}$, $E_x(x_2) = E_{x_2}, \ldots, E_x(x_n) = E_{x_n}$. We denote a finite space of states by

$$\alpha = (\{x_1, x_2, \ldots, x_n\};\ E_{x_1}, E_{x_2}, \ldots, E_{x_n})$$

and a finite space of available states by

$$\eta = (\{x_1, x_2, \ldots, x_n\};\ E_{x_1}, E_{x_2}, \ldots, E_{x_n}, S_x) \quad .$$

Note. In the set $\{x_1, x_2, \ldots, x_n\}$, the order of elements is irrelevant, but if we wrote them in some order, say, $\{x_1, x_2, \ldots, x_n\}$, then the first set E_{x_1} corresponds to x_1, that is, $E_x(x_1) = E_{x_1}$, and so on.

Example 6.1. Let $\mathbf{Z}_2 = \{0,1\}$ and $X = \{x,y,z\}$. For the space of states $(\{t,y,z\}; \mathbf{Z}_2,\mathbf{Z}_2,\mathbf{Z}_2^5)$, we have

$$E_X(t) = E_X(y) = \mathbf{Z}_2 , \quad E_X(z) = \mathbf{Z}_2^5 .$$

Let $\alpha = (\{x_1,x_2,\ldots,x_n\}; E_{x_1},E_{x_2},\ldots,E_{x_n})$ and $\beta = (\{y_1,y_2,\ldots,y_m\}; E_{y_1},E_{y_2},\ldots,E_{y_m})$ be spaces of states. A pair $(f_x,f_{\mathcal{E}_x})$ of functions

$$f_x\colon \{x_1,\ldots,x_n\} \to \{y_1,\ldots,y_m\} ,$$

$$f_{\mathcal{E}_x} : \bigcup_{x\in\{x_1,\ldots,x_n\}} E_x \to \bigcup_{y\in\{y_1,\ldots,y_m\}} E_y$$

is a morphism from α into β if and only if $E_{f_x}(y) = \beta(f_{\mathcal{E}_x})(E_y)$ for all $y \in \{x_1,\ldots,x_n\}$.

If $S_x = \prod_{y\in X} E_x(y) = \prod_{y\in X} E_y$, for a space of available states, then we will use the notation

$$\eta = (\{x_1,x_2,\ldots,x_n\}; E_{x_1} \prod E_{x_2} \prod \cdots \prod E_{x_n}) .$$

It means that $E_x(x_i) = E_{x_1},\ldots,E_x(x_n) = E_{x_n}$, and, if, $s \in S_x$, then $s(x_1) \in E_{x_1},\ldots,s(x_n) \in E_{x_n}$.

Example 6.2. For the space of available states $\eta = (\{x,y,x\}; \mathbf{Z}_2 \prod \mathbf{Z}_2 \prod \mathbf{Z}_2^3)$ (here $\mathbf{Z}_2 = \{0,1\}$), we have $E_x = E_y = \mathbf{Z}_2$, $E_z = \mathbf{Z}_2^3$ and for $s \in S_x$ we have $s(x) \in \mathbf{Z}_2$, $s(y) \in \mathbf{Z}_2$, and $s(x) \in \mathbf{Z}_2^3$.

6.2.2 For a space of (available) states $\alpha = (X,\mathcal{E}_x; E_x)$ $(\eta = (X,\mathcal{E}_x; E_x,S_x))$ the set X may be a structural set. We consider here only two cases important for applications: X is totally ordered set, and X is endowed with so-called ordered partition.

If X is a totally ordered set, that is, a structure $(X;\leq)$ and $X = \{x_1, x_2, \ldots, x_n\}$, so that $x_1 < x_2 < \ldots < x_n$, then we use the notation $X = \langle x_1, x_2, \ldots, x_n \rangle$. If $S_X = E_{x_1} \prod E_{x_2} \prod \ldots \prod E_{x_n}$, then we will use the equivalent set $E_{x_1} \times E_{x_2} \times \ldots \times E_{x_n}$, which we also denote by $\mathsf{X}_{x \in \langle x_1, \ldots, x_n \rangle} E_x$, and will use the notation

$$\eta = (\langle x_1, x_2, \ldots, x_n \rangle;\ E_{x_1} \times E_{x_2} \times \ldots \times E_{x_n})$$

$$= (\langle x_1, x_2, \ldots, x_n \rangle;\ \mathsf{X}_{x \in \langle x_1, \ldots, x_n \rangle} E_x) \quad .$$

A morphism from $\alpha = (\langle x_1, \ldots, x_n \rangle;\ E_{x_1}, \ldots, E_{x_n})$ into $\beta = (\langle y_1, \ldots, y_m \rangle;\ E_{y_1}, \ldots, E_{y_m})$ or from $\eta = (\langle x_1, \ldots, x_n \rangle;\ E_{x_1}, \ldots, E_{x_n}, S_x)$ into $\xi = (\langle y_1, \ldots, y_n \rangle;\ E_{y_1}, \ldots, E_{y_m}, S_y)$ is a pair of function $(f_x, f_{\&_x})$ which hold the conditions for morphisms in the category of spaces of (available) states, but the function f_x is increasing.

For the case $X \cap Y = \emptyset$ $(X = \langle x_1, \ldots, x_n \rangle$, $y = (\langle y_1, \ldots, y_m)$ the coproduct of α and β (η and ξ) is denoted

$$\alpha \times \beta = (\langle x_1, \ldots, x_n, y_1, \ldots, y_m \rangle;\ E_{x_1}, \ldots, E_{x_n}, E_{y_1}, \ldots, E_{y_m})$$

$$\eta \times \xi = (\langle x_1, \ldots, x_n, y_1, \ldots, y_m \rangle;\ E_{x_1}, \ldots, E_{x_n}, E_{y_1}, \ldots, E_{y_m},$$

$$S_x \bigcap S_y) \quad .$$

6.2.3 Let P_X^I be a partition of a set X. Let G define a total order on every set $X_i = P_X^I(i)$, that is, $G\colon I \to \beta(X \times X)$, so that $G(i) \subset X_i \times X_i$ and $G(i)$ is a total order on X_i. We also assume that there is the unique element $i_0 \in I$, such that X_{i_0} is unordered set. Therefore, we have the structure $(I,X;\ P_x^I, i_0, G)$, where $P_x^I \in \beta(I \times \beta(X))$, $i_0 \in I$, $G \in \beta(I \times \beta(X \times X))$. We assume that I is a finite set $I = \{i_0, i_1, \ldots, i_n\}$. In this case, we denote above structure by

$$\{x_{01}, x_{02}, \ldots, x_{0k_0}, <x_{11}, x_{12}, \ldots, x_{1k_1}>, \ldots, <x_{n1}, x_{n2}, \ldots, x_{nk_n}>\},$$

where $X_{ij} = P_x^I(i_j) = \{x_{j1}, x_{j2}, \ldots, x_{jk_j}\}$, for $j = 0,1,\ldots,n$.
We will also use the following notation

$$\{x_{01}, \ldots, x_{0k_o}, \ldots, <x_{n1}, \ldots, x_{nk_n}>\}$$

$$= \{X_{i_0}, <X_1>, \ldots, <X_p>\} = \{X_{i_0}, \mathop{<>}_{i \in I*} X_p\} \quad ,$$

where $I* = I - \{i_0\}$.

A morphism from $(I, X_1; P_{x_1}^{I_1}, i_{01}, G_1)$ into $(I_2, X_2; P_{x_2}^{I_2}, i_{02}, G_2)$ is a pair (f_I, f_x) of functions $f_I\colon I_1 \to I_2$, $f_x\colon X_1 \to X_2$ so that

$$\beta(f_I \times \beta(f_x))(P_{x_1}^{I_1}) \subset P_{x_2}^{I_2} \ , \ i_{02} = f_I(i_{01}) \ ,$$

$$\beta(f_I \times \beta(f_x \times f_x))(G_1) \subset G_2 \quad ,$$

that is, the diagrams

$$\begin{array}{ccc} I_1 & \xrightarrow{f_I} & I_2 \\ {\scriptstyle P_{x_1}^{I_1}} \downarrow & & \downarrow {\scriptstyle P_{x_2}^{I_2}} \\ \beta(X_1) & \xrightarrow{\beta(f_x)} & \beta(X_2) \end{array} \quad ,$$

$$\begin{array}{ccc} I_1 & \xrightarrow{f_I} & I_2 \\ G_1 \big\downarrow & & \big\downarrow G_2 \\ \beta(X_1 \times X_1) & \xrightarrow{\beta(f_x \times f_x)} & \beta(X_2 \times X_2) \end{array}$$

are communtative and $i_{02} = f_I(i_{01})$. Therefore, if $x \in X_i$, $y \in X_i$, and $x \leq_{G_i} y$, then $f_x(x) \leq G_{f_I(i)} f_x(y)$, where

$$x \leq_{G_i} y \iff (x,y) \in G_i,\ u \leq_{G_{f_I(i)}} w \iff (u,w) \in G_{f_I(i)}.$$

The structure $(I,X;\ P_x^I, i_0, G) = \{X_{i0}, \langle\,\rangle_{i \in I*} X_i\}$ is called an *ordered partition*. Given two ordered partitions

$(I_1, X;\ P_x^I, i_{01}, G_{I*}) = \{X_{i_{01}}, \langle\,\rangle_{i \in I_{1*}} X_i\}$ and

$(I_2, Y;\ P_x^{I_2}, i_{02}, G_{I_2}) = \{Y_{i_{02}}, \langle\,\rangle_{i \in I_{2*}} Y_i\}$ so that $X \cap Y = \emptyset$.

$I_I \cap I_2 = \emptyset$. The ordered partitions

$$(\{i_0\} \cup I_{1*} \cup I_{2*},\ X \cup Y;\ P_{x_1 \cup x_2}^{I_1 \cup I_2},\ i_0,\ G_{I_1 \cup I_2})$$

$$= \{X_{i_{01}} \cup X_{i_{02}},\ \langle\,\rangle_{i \in I_{1*}} X_i,\ \langle\,\rangle_{i \in I_{2*}} Y_i\} \quad ,$$

where

$$P_{x_1 \cup x_2}^{I_1 \cup I_2}(i) = \begin{cases} X_{i_{01}} \cup X_{i_0} & \text{if } i = i_0 \\ X_i & \text{if } i \in I_{1*} \\ Y_i & \text{if } i \in I_{2*} \end{cases} \quad ,$$

$$G_{I_1 \cup I_2}(i) = \begin{cases} G_{I_1}(i) & \text{if } i \in I_{1*} \\ G_{I_2}(i) & \text{if } i \in I_{2*} \end{cases}$$

and the pair (σ_1,σ_2) of morphisms $\sigma_1 = (f_{I_1}, f_x)$, $\sigma_2 = (f_{I_2}, f_y)$ such that

$$f_{I_1}: I_1 \to \{i_0\} \cup I_{1*} \cup I_{2*}, \quad f_{I_1}(i_{01}) = i_0, \quad f_{I_1}(i) = i$$

$$f_{I_2}: I_2 \to \{i_0\} \cup I_{1*} \cup I_{2*}, \quad f_{I_2}(i_{02}) = i_0, \quad f_{I_2}(i) = i$$

$$f_x: X \to X \cup Y, \quad f_x(t) = t$$

$$f_y: Y \to X \cup Y, \quad f_y(t) = t$$

comprise a coproduct of the given ordered partition.

Defintion 6.2. A space of (available) states for ordered partition X is a structure $(X,I,\&_x;\ P_x^I, E_x, i_0, G)$ $((X,I,\&_x;\ P_x^I, E_x, i_0, G, S_x))$, where $(I,X;P_x^I, i_0, G)$ is an ordered partition and $(X,\&_x;\ E_x)$ $((X,\&_x;\ E_x,S_x))$ is a space of (available) states for X.

We can define a space of (available) states for ordered partition X is a structure $(\hat{X},\&_x;\ E_x)$ $((\hat{X},\&_x;\ E_x,S_x))$ where $\hat{X} = (I,X;\ P_x^I, i_0, G)$ is an object of the category of ordered partitions.

If $I = \{i_0, i_1, \ldots, i_n\}$, then a space of states α and available states η are denoted as

$$\alpha = (\{x_{01},\ldots,x_{0k_0},\ldots,<x_{n1},\ldots,x_{nk_n}>\};$$

$$E_{x_{01}},\ldots,E_{x_{0k_0}},\ldots,E_{x_n},\ldots,E_{x_{nk_n}}) \quad ,$$

$$\eta = (\{x_{01},\ldots,x_{0k_0},\ldots,<x_{n1},\ldots,x_{nk_n}>\};$$
$$E_{01},\ldots,E_{x_{0k_0}},\ldots,E_{x_{n1}},\ldots,E_{x_{nk_n}},S_x) \quad .$$

Here

$$S_x \subset E_{x_{01}} \prod E_{x_{02}} \prod \cdots \prod E_{x_{0k_0}} \prod (E_{x_{11}} \times \cdots \times E_{x_{1k_1}})$$
$$\prod \cdots \prod (E_{x_{n1}} \times \cdots \times E_{x_{nk_n}}) \quad . \tag{*}$$

If S_x equals the right hand side expression in (*), then we use the following notation

$$\eta = (\{x_{01},\ldots,x_{0k_0},\ldots,<x_{n_1},\ldots,x_{nk_n}>\};$$
$$E_{x_0} \prod \cdots \prod E_{x_{0k_0}} \prod \cdots \prod (E_{x_{n1}} \times \cdots \times E_{x_{nk_n}})) \quad .$$

Example 6.3. Given $X = \{x,y,<z,t>,<u,w>\}$ and $E_x = Z_2^2$, $E_y = E_z = E_u = E_w = X$ $(X = \{0,1,2,\ldots,9,A,B,C,D,E,F\})$, $E_t = Z_2^3$. The space of states for X is

$$\alpha = (\{x,y,<z,t>,<u,w>\};\ Z_2^2, X, X, Z_2^3, X, X)$$

and the space of available states

$$\eta = (\{x,y,<z,t>,<u,w>\};\ Z_2^2 \prod X \prod (X \times Z_2^3) \prod (X \times X)).$$

A morphism from $\alpha_1 = (X_1,I_1,\&_{x_1};P_{x_1}^{I_1},E_{x_1},i_{01},G_1)$ (from $\eta_1 = (X_1,I_1,\&_{x_1};P_{x_1}^{I_1},E_{x_1},i_{01},G_1,S_{x_1}))$ into

$\alpha_2 = (X_2, I_2, \mathcal{E}_{X_2}; P^{I_2}_{X_1}, E_{X_2}, i_{02}, G_2)$ (into $\eta_i = (X_2, I_2, \mathcal{E}_{X_2};$ $P^{I_2}_{X_2}, E_{X_2}, i_{02}, G_2, S_{X_2}))$ is a triple $(f_I, f_X, f_{\mathcal{E}_X})$, where $f_I\colon I_1 \to I_2$, $f_X\colon X_1 \to X_2$, $f_{\mathcal{E}_X}\colon \mathcal{E}_{X_1} \to \mathcal{E}_{X_2}$, such that (f_I, f_X) is a morphism in the category of ordered partitions and $(f_X, f_{\mathcal{E}_X})$ is a morphism in the category of spaces of (available) states.

For the spaces of available states η_1 and η_2 if $X_1 \cap X_2 = \emptyset$, $I_1 \cap I_2 = \emptyset$, the coproduct will be denoted as $\{\eta_1, \eta_2\}$. We have

$$\{\eta_1, \eta_2\} = (X_1 \cup X_2,\ I_1 \cup I_2,\ \mathcal{E}_{X_1} \coprod \mathcal{E}_{X_2};\ P^{I_1 \cup I_2}_{X_1 \cup X_2},\ E_{X_1 \cup X_2},\ G_{P_1 \cup P_2},\ S_{X_1 \frown X_2}) .$$

Therefore, if

$$\eta_1 = \left(\{X_{i_{01}}, \langle\ \rangle_{i \in I_{1*}} X_i\};\ \prod_{x \in X_{i_{01}}} E_x \prod_{x \in \langle\ \rangle_{i \in I_{1*}} X_i} \left(\bigtimes_{x \in X} E_x\right)\right)$$

$$\eta_2 = \left(\{X_{i_{02}}, \langle\ \rangle_{i \in I_{2*}} X_i\};\ \prod_{x \in X_{i_{02}}} E_x \prod_{x \in \langle\ \rangle_{i \in I_{2*}} X_i} \left(\bigtimes_{x \in X} E_x\right)\right)$$

and if $X_{i_{01}} \cup X_{i_{02}} = X_{i_0}$, $I^* = I_{1*} \cup I_{2*}$, then

$$\{\eta_1, \eta_2\} = \left(\{X_{i_0}, \langle\ \rangle_{i \in I^*} X_i\};\ \prod_{x \in X_{i_0}} E_x \prod_{x \in \langle\ \rangle_{i \in I^*} X_i} \left(\bigtimes_{x \in X} E_x\right)\right) .$$

We used the notation

$$\prod_{x\in X_{i_0}} E_x \prod_{\substack{x\in \langle\ \rangle X_i \\ i\in I*}} \left(\bigtimes_{x\in X} E_x \right)$$

$$= E_{x_0} \prod \cdots \prod E_{x_{0k_0}} \prod (E_{11} \times \cdots \times E_{1k_1})$$

$$\prod \cdots \prod (E_{n1} \times \cdots \times E_{nk_n}) \quad .$$

Example 6.4. Given η_1 and η_2

$$\eta_1 = \left(\{x,\langle y,z\rangle\};\ Z_2 \prod (Z_2 \times X)\right)$$

$$\eta_2 = \left(\{t,u,\langle w,s\rangle\};\ X^2 \prod X \prod (Z_2 \times Z_2^2)\right) \quad .$$

Then

$$\{\eta_1,\eta_2\} = \left(\{x,t,u,\langle y,z\rangle,\langle w,s\rangle\};\right.$$

$$\left. Z_2 \prod X^2 \prod X \prod (Z_2 \times X) \prod (Z_2 \times Z_2^2)\right) \quad .$$

6.3 Addition Instruction of IBM/370 Computer

6.3.1 In this point, we apply the technique described in points 6.1 and 6.2 to modeling the addition instruction of IBM/370 computer. Every instruction of a computer is an operation on the space of available states of a subsystem of the computer. Therefore, the set of programs in a machine language describes the algebraic object more complex than universal algebra. Namely, any program being an operation is related to a subsystem of the computer. It seems, that the language of category theory is convenient for description of this object. In this point, we consider only one instruction of the specific computer IBM/370. We would like to show how the technique developed in points 6.1 and 6.2 can be used effectively for modeling computers and their languages.

We consider the instruction "Add Registers" (AR) in the case when the computer is in the operating state and in the basic-control (BC) mode. The AR instruction adds a fullword binary integer stored in the register R2 to a fullword binary integer stored in the register R1 and places the result into the register R1. If the overflow takes place, the program interruption may occur. To describe the instruction, we should consider the following registers: the current program status word (CPSW), the new program status word (NPSW), the old program status word (OPSW), the general register R1, and the general register R2. Therefore, we should consider the subsystem (the space of available states) called AR and denoted by η_{AR}. This subsystem is a coproduct of registers CPSW (the space of available states η_{CPSW}), NPSW (the space of available states η_{NPSW}), OPSW (the space of available states η_{OPSW}), R1 (the space of available states η_{R1}), and R2 (the space of available states η_{R2}). Therefore,

$$\eta_{AR} = \{\eta_{CPSW}, \eta_{NPSW}, \eta_{OPSW}, \eta_{R1}, \eta_{R2}\} \quad .$$

We describe now the structure of a program status word regiter (the space of available states η_{PSW}) and a general register (the space of available states η_R).

A program status word register consists of nine subregisters:

sm is the system-mask register,

Key is the PSW-Key register,

e is the extended-control-mode register,

m is the machine-check register,

w is the wait-state register,

p is the problem-state register,

ic is the interruption-control register,

pc is the program-control register,

ad is the instruction-address register.

The space of available states of PSW depends on the state of e register, which may be in the two states denoted by 0 and 1, that is, $E_e = \{0,1\} = \mathbf{Z}_2$. We assume, that e is in the state 0; in this case the system is called to be in the basic-control (BC) mode.

The system-mask register consists of the following sub-registers:

cm is the channel-mask register

io is the input/output (I/O) mask register,

ex is the external-mask register.

The space of states of cm is $E_{cm} = \mathbf{Z}_2^6$. The I/O mask register io and the external-mask register may be in one of the states from $\mathbf{Z}_2$, that is $E_{io} = E_{ex} = \mathbf{Z}_2$. Therefore, the system mask register is the following space of available states

$$\eta_{sm} = (\langle cm,io,ex\rangle;\ \mathbf{Z}_2^6 \times \mathbf{Z}_2 \times \mathbf{Z}_2).$$

The space of states of the PSW-Key register is $\mathbf{X} = \{0,1,2,\ldots,9,A,B,C,D,E,F\}$, that is, $E_{Key} = \mathbf{X}$. Therefore, $\eta_{Key} = (\{Key\};\mathbf{X})$.

The registers extended-control mode e, machine-check-mask m, wait-state w, and problem-state p may be in one of the states from $\mathbf{Z}_2$, that is, $E_e = E_m = E_w = E_p = \mathbf{Z}_2$. We assume that the state of e is 0. Therefore

$$\eta_e = (\{e\};\mathbf{Z}_2)\ ,\quad \eta_m = (\{m\};\mathbf{Z}_2)\ ,\quad \eta_w = (\{w\};\mathbf{Z}_2)\ ,$$
$$\eta_p = (\{p\};\mathbf{Z}_2)\quad .$$

The register ic describes an interruption condition. The set of all interruption conditions is denoted by IC. Therefore, $E_{ic} = IC$, and $\eta_{ic} = (\{ic\}; IC)$.

The program-control register consists of the following subregisters:

iℓc is the interruption-length-code register,

cc is the condition-code register,

pm is the program-mask register.

The space of states of the registers iℓc and cc is the same and consists of words (strings) of two digits from Z_2. Therefore, $E_{i\ell c} = E_{cc} = Z_2^{*2}$ (X^{*n} means the set of words of the length n in the alphabet X). The space of states of pm is $E_{pm} = Z_2^4$. Therefore,

$$\eta_{pc} = (\langle i\ell c, cc, pm\rangle;\ Z_2^{*2} \times Z_2^{*2} \times Z_2^4) \quad .$$

A state of the instruction address is a word of length 6 in the alphabet X. Therefore, $\eta_{ad} = (\{ad\}; X^{*6})$. Therefore, the program-status word register is the following coproduct of the spaces of available states

$$\eta_{PSW} = \eta_{sm} \times \eta_e \times \eta_m \times \eta_w \times \eta_p \times \eta_{ic} \times \eta_{pc} \times \eta_{ad} \quad .$$

It is also convenient to describe this space of available states in the PL/1 language by using the declaration statement and by extension of the data definition.

```
DCL   1 PSW,
        2 SM,                    /*SYSTEM-MASK*/
          3 CM(6)   BIT(1),      /*CHANNEL-MASK*/
          3 IO      BIT(1),      /*INPUT/OUTPUT-MASK*/
          3 EX      BIT(1),      /*EXTERNAL-MASK*/
        2 KEY       HEX(1),      /*PSW-KEY*/
        2 E         BIT(1),      /*EXTENDED-CONTROL-MODE*/
        2 M         BIT(1),      /*MACHINE-CHECK*/
        2 W         BIT(1),      /*WAIT-STATE*/
        2 P         BIT(1),      /*PROBLEM-STATE*/
        2 IC      DATA(IC),      /*INTERRUPTION-CONTROL*/
        2 PC,                    /*PROGRAM-CONTROL*/
          3 ILC     BIT(2),      /*INTERRUPTION-LENGTH-CODE*/
          3 CC      BIT(2),      /*CONDITION-CODE*/
          3 PM(4)   BIT(1),      /*PROGRAM-MASK*/
        2 AD        HEX(6);      /*INSTRUCTION-ADDRESS*/
```

A general register is described by its number (NUM) and the content (CONT) with an array of thirty-two bits. Therefore, the space of states for NUM $E_{NUM} = X$, and for CONT $E_{CONT} = Z_2^{32}$. Hence

$$\eta_R = (\{NUM, CONT\};\ X \prod Z_2^{32}) \quad .$$

By using the extension of PL/1, we have

```
DCL   1 REG,
        2 NUM        HEX(1),   /*NUMBER OF REGISTER*/
        2 CONT(32)   BIT(1);   /*CONTENTS OF REGISTER*/
```

We can describe now in the extended PL/1 the subsystem of the computer associated with AR instruction. We use the name AR_SUB.

```
DCL    1 AR_SUB,
         2 CPSW   LIKE PSW,   /*CURRENT PSW*/
         2 NPSW   LIKE PSW,   /*NEW PSW*/
         2 OPSW   LIKE PSW,   /*OLD PSW*/
         2 R1     LIKE REG,   /*REGISTER 1*/
         2 R2     LIKE REG;   /*REGISTER 2*/
```

6.3.2 To describe the instruction AR as an operation on the space of available states η_{AR}, we consider, first, the addition operation on the set of fullword binary integers. On the set $\mathsf{X} = \{0,1,2,\ldots,9,A,B,C,D,E,F\}$, we have the following "addition" operation

$$\oplus_X : \mathsf{X} \times \mathsf{X} \to \mathsf{X}^{*2} : (x,y) \to x \oplus_X y$$

described by the following table.

(See following page for the table.)

The set of fullword binary integers is a totally ordered set which consists of strings of eight digits from X. We denote this set by Z_F. We have

$$\mathsf{Z}_F = \langle 80000000, 80000001, \ldots, FFFFFFFF, 00000000, 00000001, \ldots, 7FFFFFFF\rangle \quad .$$

Thus, for every element $z \in \mathsf{Z}_F$

$$z \in \mathsf{Z}_F \iff (z = x_1x_2x_3x_4x_5x_6x_7x_8 \text{ and } x_i \in \mathsf{X} \text{ for } i = 1,2,\ldots,8) \quad .$$

The element 80000000 is the minimal element, the element 7FFFFFFF is the maximal element. Elements 80000000,..., FFFFFFFF are called negative elements, elements 00000001, ...,7FFFFFFF are called positive elements.

x \ y	0	1	2	3	4	5	6	7	8	9	A	B	C	D	E	F
0	00	01	02	03	04	05	06	07	08	09	0A	0B	0C	0D	0E	0F
1	01	02	03	04	05	06	07	08	09	0A	0B	0C	0D	0E	0F	10
2	02	03	04	05	06	07	08	09	0A	0B	0C	0D	0E	0F	10	11
3	03	04	05	06	07	08	09	0A	0B	0C	0D	0E	0F	10	11	12
4	04	05	06	07	08	09	0A	0B	0C	0D	0E	0F	10	11	12	13
5	05	06	07	08	09	0A	0B	0C	0D	0E	0F	10	11	12	13	14
6	06	07	08	09	0A	0B	0C	0D	0E	0F	10	11	12	13	14	15
7	07	08	09	0A	0B	0C	0D	0E	0F	10	11	12	13	14	15	16
8	08	09	0A	0B	0C	0D	0E	0F	10	11	12	13	14	15	16	17
9	09	0A	0B	0C	0D	0E	0F	10	11	12	13	14	15	16	17	18
A	0A	0B	0C	0D	0E	0F	10	11	12	13	14	15	16	17	18	19
B	0B	0C	0D	0E	0F	10	11	12	13	14	15	16	17	18	19	1A
C	0C	0D	0E	0F	10	11	12	13	14	15	16	17	18	19	1A	1B
D	0D	0E	0F	10	11	12	13	14	15	16	17	18	19	1A	1B	1C
E	0E	0F	10	11	12	13	14	15	16	17	18	19	1A	1B	1C	1D
F	0F	10	11	12	13	14	15	16	17	18	19	1A	1B	1C	1D	1E

The "addition" operation $\oplus_F$ on the set Z_F is defined as follows

$$\oplus_F : Z_F \times Z_F \to Z_F \times \{\text{correct, overflow}\} : (x,y) \to x \oplus_F y .$$

If $x = X_1X_2X_3X_4X_5X_6X_7X_8$ and $y = Y_1Y_2Y_3Y_4Y_5Y_6Y_7Y_8$, then $x \oplus_F y = (Z_1Z_2Z_3Z_4Z_5Z_6Z_7Z_8, \text{cond})$, where cond $\in$ {correct, overflow}.

$$\hat{Z}_8^1 = X_8 \oplus_X Y_8 , \quad \hat{Z}_8 = \hat{Z}_8^1 , \quad C_8 = pr_1(\hat{Z}_8) , \quad \text{and}$$

$$Z_8 = pr_2(\hat{Z}_8);$$

$$\hat{Z}_7^1 = X_7 \oplus_X Y_7 , \quad \hat{Z}_7 = pr_2(\hat{Z}_7) \oplus_X C_8 ,$$

$$C_7 = \max(pr_1(\hat{Z}_7'), pr_1(\hat{Z}_7)) , \quad \text{and} \quad Z_7 = pr_2(\hat{Z}_7);$$

... ...

$$\hat{Z}_i' = X_i \oplus_X Y_i , \quad \hat{Z}_i = pr_2(\hat{Z}_i') \oplus_X C_{i+1} ,$$

$$C_i = \max(pr_1(Z_i'), pr_1(\hat{Z}_i)) , \text{ and } \quad Z_i = pr_2(\hat{Z}_i)$$

for $i = 1,2,3,\ldots,7$.

cond = correct

if

i) $X_1 \leq 7$, $Y_1 \leq 7$, and $Z_1 \leq 7$;

ii) $X_1 \geq 8$, $Y_1 \geq 8$, and $Z_1 \geq 8$;

iii) x and y have different signs, that is, if one is positive, then the other is negative.

otherwise

cond = overflow .

By using the above algorithm, it is not difficult to write a subroutine ADD_FW in PL/1, which accepts the contents of two registers R1 and R2, interprets them as two fullword binary integers, adds them and puts the result in register R1. If the result is correct, then the subroutine assigns the value CORR (correct) to the parameter COND, otherwise it assigns the value OVFL (overflow) to the parameter COND. The subroutine also sets the CC as follows: CC = '00'B if the result equals 00000000, CC = '01'B if the result is negative, CC = '10'B if the result is positive, and CC = '11'B if COND = OVFL. We denote this subroutine by ADD_FW(R1,R2,COND,CC).

The second functional subroutine we are going to use is ADD_AD(AD1,AD2). This subroutine returns the sum of two addresses AD1 and AD2. An address is a word of six hexadecimal digits. The subroutine extends them on the left to eight hexadecimal digits and then adds them as fullword binary integers. The six last hexidecimal digits of the result is the new address. Because the representation of the operation specified by the AR instruction in mathematical terms is very cumbersome, we represent this operation in the PL/1 language.

```
ADD_RR:   PROC(AR_SUB);
          DCL  COND  CHAR(4),
               CC    BIT(2);

/*IF CPU IS IN WAIT STATE, THEN DO NOTHING AND RETURN*/
          IF CPSW.W = '1'B THEN RETURN;
          CALL ADD_FW(R1,R2,COND,CC);
          CPSW.AD = ADD_FW(CPSW.AD, X'000002');

/*THE LENGTH OF THE INSTRUCTION AR EQUALS 2; ILC = '01'B*/
          CPSW.CC = CC;
          IF(COND = OVFL) & (PM(1) = '1'B) THEN DO;
          CPSW.IC = X'08';

/*WE USED THE VALUE X'08' FOR IC FOR FIXED-POINT OVERFLOW*/
          CPSW.ILC = '01'B;
          OPSW = CPSW; CPSW = NPSW; RETURN;
                                             END;
END AD_RR;
```

Bibiliography

1. N. Bourbaki, *Theory of Sets, Elements of Mathematics*, Addison-Wesley, Reading, Mass., 1968.
2. A. Tuzhilin, *Category Theory of Structural Sets with Applications to Mathematical Modelling and Systems Analysis*, An International Journal of Mathematical Modelling (in print).
3. A. Tuzhilin, *Category of Spaces of Available States with Application to Modelling of Computer Languages*, Proceedings of Fourth International Conference on Mathematical Modelling, Zurich, 1983.

4. M. Tausner and A. Tuzhilin, *Modelling the Apple II Using Categories of Spaces of Available States*, Proceedings of Fourth International Conference on Mathematical Modelling, Zurich, 1983.

Department of Computer Science, College of Staten Island, City University of New York, Staten Island, New York, 10301

ABCDEFGHIJ–AMS/AP–89876